Scientific Objectivity

Adopting a transcendental approach, the book addresses debates on "scientific objectivity" within the contemporary philosophy of science.

The author explores the multifaceted meanings and ongoing debates surrounding objectivity and introduces publicity as a necessary condition for scientific knowledge. In addition, he discusses the various ways in which scientific claims can be considered objective. The author highlights its role in ensuring the essential conditions for communication and mutual understanding that are central to scientific practice. Through an analysis of intersubjectivity in Hermann Weyl's spacetime theory, the book provides an application of this framework, illustrating how objectivity emerges through shared rational engagement. By integrating these arguments, the book offers a profound reexamination of scientific knowledge, objectivity, and the social dimensions of rational inquiry.

This book will appeal to students and scholars of the philosophy, ethics, and sociology of science.

Dong Luo is an associate professor at the Institute for Advanced Studies in Philosophy, Science, and Technology at the South China University of Technology. He has research interests in the history and philosophy of mathematics and physics, philosophy of science, and comparative philosophy.

Scientific Objectivity

A Transcendental Approach

Dong Luo

Routledge
Taylor & Francis Group
LONDON AND NEW YORK

First published 2026
by Routledge
4 Park Square, Milton Park, Abingdon, Oxon OX14 4RN

and by Routledge
605 Third Avenue, New York, NY 10158

Routledge is an imprint of the Taylor & Francis Group, an informa business

British Library Cataloguing in Publication Data
A catalogue record for this book is available from the British Library

ISBN: 978-1-041-14060-3 (hbk)
ISBN: 978-1-041-32650-2 (pbk)
ISBN: 978-1-003-78378-7 (ebk)

DOI: 10.4324/9781003783787

Typeset in Times New Roman
by KnowledgeWorks Global Ltd.

Contents

Figures and Tables

Figures

Tables

Acknowledgements

The research presented in this book has a history, originating in my investigations into the epistemological roles of symmetry in physical theories roughly ten years ago. A pivotal moment came with my attempt to interpret Hermann Weyl's view on objectivity and invariance, an effort that played a foundational role in the early stages of this project and is largely reflected in Chapter 5.

Initially, I struggled to situate Weyl's claim within his own philosophical and mathematical framework. In seeking to make sense of it, I gradually introduced concepts and perspectives that extended well beyond Weyl's original framework. This brought the work into a somewhat uneasy state, caught between the effort to understand Weyl and the emergence of my own philosophical standpoint, particularly since my engagement with various notions of objectivity bore little direct connection to the study on Weyl. Over time, these conceptual additions transformed the project so significantly that it diverged markedly from its original aims.

In 2023, I began approaching the problem of understanding from the perspective of intersubjectivity. This enabled me to articulate the dimension of referential identity within the problem of publicity through the lens of understanding itself. That insight once again redirected the trajectory of the project onto a path significantly divergent from the one I had initially envisioned.

I am especially grateful to Dennis Dieks for encouraging me to study Hermann Weyl and for the extensive and illuminating discussions on Weyl and the central themes of this book; to Dingcheng Ren and Jan Sprenger for the early discussions on scientific objectivity; and to Steve Fuller for the discussions on the problems of understanding and misunderstanding.

Also, I am grateful to the generative AI model Qwen3-Max for the iterative exchanges, which helped me articulate my ideas with increasing clarity, although some points may still benefit from further refinement, and for its assistance in refining the language and improving the clarity of early drafts of this manuscript. The ideas, arguments, and research content are entirely my own. Responsibility for all claims, interpretations, and any remaining errors rests solely with me.

Dong Luo
Guangzhou
December 2025

Introduction

Objectivity has long been regarded as a hallmark of scientific inquiry. But what, exactly, is scientific objectivity? It is tempting to take it as meaning that scientific knowledge must be justifiable independently of anybody's whims. In this sense, objectivity aligns with Kant's notion of rational universality: if something is valid for *everyone*, provided that they have reason, then its grounds are objective and sufficient[1]. Scientific objectivity in this sense promises that scientific claims are universally testable and intelligible, anchored in a framework of collective verification that transcends individual subjectivity, biases, interests of specific communities, etc.

However, scientific objectivity in this sense imposes a remarkably strong requirement. On one hand, in practice, it is impossible for any statement to be *actually* tested by *everyone*. If universal testing is unattainable, how can we legitimately claim that a statement *is* justified by any rational agent and thereby assert that it constitutes "objective knowledge"? Moreover, even if all current testers agree on a statement, such consensus offers no guarantee that future investigators will concur, underlying a problem of induction. The very idea that a claim is justifiable by *anyone*, now and in the future, rests on an inductive leap that cannot itself be secured. On the other hand, there are theories (e.g., Thomas Kuhn's theory of scientific revolutions and paradigm shifts) challenging the very idea that scientific statements could be valid or agreed upon by anybody. They argued that the criteria for scientific evaluation are contingent on the perspectives of the scientists. And it is even unnecessary to imagine the existence of "one full, objective, true account of nature" (Kuhn 1962, 171)[2]. For example, the standards of objectivity in Newtonian physics, which prioritized deterministic laws, differ markedly from those in quantum mechanics, where probabilistic interpretations prevail. This historical dependence seems to suggest that the criterion for objectivity of scientific theories is not a fixed ideal but a context-dependent construct shaped by the prevailing scientific framework.

Moreover, the concept of "objectivity" in science is burdened by a profusion of definitions that often carry contradictory meanings. Over a dozen of distinct senses of objectivity have been identified[3], with some emphasizing

DOI: 10.4324/9781003783787-1

epistemic efforts to transcend individual perspectives, aiming for "a view from nowhere" (Nagel 1986)[4], while others equate objectivity with freedom from human intervention, such as following mechanical protocols or automatically recording reality; and some insist on complete insulation from values (i.e., value-freedom), whereas yet another strand accepts that values are inevitable and instead advocates for a balanced or transparent incorporation of them (i.e., value-balanced approaches)[5].

Further complicating the picture, the concept of objectivity has also evolved through distinct epistemic virtues across the history of science. Lorraine Daston and Peter Galison, in their landmark study *Objectivity* (2007)[6], describe a major shift in the image of objectivity: from "truth-to-nature" before the nineteenth century, which emphasized capturing essential, idealized forms of natural phenomena; to mechanical objectivity in the nineteenth century, which sought to eliminate human intervention through technologies like photography, aiming for unmediated, automatic records of reality; to "trained judgment" after the early twentieth century, which is grounded in the consensus within specialized communities, cultivated through years of shared training and experience. These historical transformations reveal that objectivity is not a timeless epistemic anchor but a historically contingent ideal with criteria shifting alongside scientific practices, instruments, and institutional norms. Moreover, recent studies further argue that objectivity cannot be understood apart from its social, political, ethical dimensions (e.g., Kitcher 1993; Longino 1990)[7]. Also, some feminists even argue that traditional ideals of objectivity often mask power imbalances and epistemic exclusions (e.g., Lloyd 1995, 2005)[8].

Because of the entanglement of the notions of objectivity and the complexity of its conceptual landscape, numerous researchers have proposed classifications of the various senses of objectivity (e.g., Axtell 2016; Megill 1994; Mühlholzer 1988)[9], hoping at least to clarify the picture within scientific discourse. However, perhaps disappointingly, these taxonomic efforts reveal that the concept of scientific objectivity possesses what Heather Douglas (2004) calls an "irreducible complexity" and resists reduction to any single essential meaning.

The proliferation of notions of "objectivity" and the difficulty of defining a universal meaning for scientific objectivity have led some philosophers to propose abandoning the concept of objectivity altogether. Ian Hacking, for instance, argues that if one wishes to discuss the desirable features or epistemic virtues of scientific practice, such as trustworthiness or reliability, she can do so directly, by referring to the specific norms, practices, or qualities that underwrite those virtues, without recourse to the overloaded and ambiguous concept of objectivity. Hence, "Let's not talk about objectivity" (Hacking 2015)[10].

Despite these difficulties, frustrations, and pessimism in understanding "objectivity", some philosophers still try in some way to "save" the concept.

For example, Inkeri Koskinen's (2021)[11] proposes a "risk account of scientific objectivity" that aims to unify the diverse, context-sensitive senses of objectivity found in contemporary literature under a single, essentially descriptive framework by the notion of reliance: to characterize something as objective is to assert its reliability as a basis for belief or action across diverse epistemic agents. Another example is Nancy Cartwright et al. (2022)[12], who argue that notions of objectivity play dual roles as both "world-guided" and "action-guiding": they describe features of reality while also informing how we ought to act in response to it. On this view, we should not discard the concept of objectivity altogether. Rather, by studying its varied uses, we come to see objectivity as something uncovered through careful, situated judgment—one that is responsive to the local interplay of epistemic needs, ethical constraints, and practical realities.

However, these few efforts are far from successfully saving the concept of scientific objectivity from its current crisis, especially in an era marked by post-truth discourse and the various scandals that have eroded public trust in scientific communities. Not to mention that objectivity carries a rich philosophical history that cannot be reduced to mere reliance or trust.

In this book, I propose a framework for understanding scientific objectivity through a transcendental approach, one that situates the concept of scientific objectivity within the broader history of the problem of objectivity in general philosophy. My aim is neither to defend a single, universal conception of objectivity, nor to abandon the concept altogether. Rather, I argue that the pursuit of objectivity can and should remain equitable for its merit in fostering mutual understanding in scientific practice.

Central to my approach is the notion of publicity: the idea that scientific claims must, at a minimum, be accessible to and evaluable by multiple researchers. I treat publicity not as a sufficient guarantee of truth, but as a necessary condition for objectivity—precisely why I regard my approach as transcendental. On this view, objectivity is not a fixed or timeless standard, but an ideal and dynamic process that guides scientists toward open communication, mutual accountability, and the ongoing achievement of common ground among individuals and even across diverse epistemic communities.

The first chapter of this book explores the multifaceted nature of scientific objectivity and surveys various attempts to classify its different senses. These explorations reveal the significant challenges and the likely impossibility of articulating a single, universal conception of objectivity that could command consensus across all scientific and philosophical contexts.

Rather than seeking to defend a general or unified notion of scientific objectivity, the second chapter proposes a minimal, necessary condition for objectivity in science: publicity. I argue that, regardless of how one interprets scientific objectivity, whether as fidelity to the world "as it is," as freedom from personal bias, as procedural discipline, or as intersubjective agreement, any viable account must require that scientific representations constitute public knowledge: knowledge that is, in principle, accessible to and intelligible by multiple rational agents, rather than remaining the private possession of any individual. While *actual* agreement among scientists may often remain elusive, the very idea of objectivity aspires such agreement, and the aspiration rests on the public nature of scientific inquiry.

Taking publicity as this necessary condition, the third chapter examines the epistemological gap between public knowledge and the immediate, private experiences of individual subjects. Bridging this gap connects the debate over scientific objectivity in the philosophy of science to a long-standing problem in traditional philosophy: the problem of objectivity as it arises in discussions of other minds, intersubjectivity, and the constitution of a shared world. Thus, the issue of publicity in this book serves as a link between contemporary concerns in philosophy of science and classical philosophy inquiries into the conditions of objective experience.

In the fourth chapter, building on this framework, I develop a transcendental approach to the constitution of publicity in scientific representation. Beginning from the irreducible starting point of private, subjective experience, I argue that individuals must adopt a specific epistemic stance: to transform their private representations into genuinely public scientific knowledge, they must go beyond their own immediate perspectives and actively strive to understand the possible experiences and viewpoints of others—imagining, as it were, what the world would look like "from their shoes." This act of epistemic compromise, "committing not only to one's own experience but also to the effort of grasping others," creates the conditions for mutual understanding and potential agreement. While such an agreement might not always be sufficiently guaranteed, its possibility can emerge through communications among individuals who jointly commit to this strategy of perspective-taking.

The fifth chapter offers a concrete case study of this framework through an analysis of intersubjectivity in Hermann Weyl's spacetime theory. Weyl's introduction of gauge symmetry into his unified field theory provides an illustration of how publicity, as a necessary condition for objectivity, might be actively constituted within scientific practice.

The Conclusion concludes the book by synthesizing these arguments and reflecting on their implications for our understanding of scientific knowledge, objectivity, and the social dimensions of rational inquiry.

Notes

1 See Kant, Immanuel. *Critique of Pure Reason*, A820/B848 (Translated by Partricia Kitcher. Indianapolis/Cambridge: Hackett, 1996): "If the assent is valid for everyone, provide that he has reason, then its basis is sufficient *objectively*, and the assent is then called *conviction*."

2 Kuhn, Thomas. *The Structure of Scientific Revolutions*. Chicago: University of Chicago Press, 1962.

3 Janack, Marianne. "Dilemmas of objectivity." *Social Epistemology*, 16 (November, 2002): 267–281.
4 Nagel, Thomas. *The View From Nowhere*. New York: Oxford University Press, 1986.
5 Chapter 1 will present a more detailed discussion on the complexity of the concept of objectivity.
6 Daston, Lorraine, and Peter Galison. *Objectivity*. Cambridge, MA: MIT Press, 2007.
7 Longino, Helen. *Science as Social Knowledge*. Princeton, NJ: Princeton University Press, 1990. Kitcher, Philip. *The Advancement of Science: Science Without Legend and Objectivity Without Illusions*. New York: Oxford University Press, 1993.
8 Lloyd, Elisabeth. "Objectivity and the double standard for feminist epistemologies." *Synthese*, 104 (September, 1995), 351–381. Lloyd, Elisabeth. *The Case of the Female Orgasm: Bias in the Science of Evolution*. Cambridge, MA: Harvard University Press, 2005.
9 Mühlholzer, Felix. "On objectivity." *Erkenntnis*, 28 (1988): 185–230; Megill, Allan. "Four senses of objectivity." In *Rethinking Objectivity*. Edited by Allan Megill, 1–29. Durham: Duke University Press, 1994; Axtell, Guy, *Objectivity*. Cambridge: Polity Press, 2016.
10 Hacking, Ian. "Let's not talk about objectivity." In *Objectivity in Science: New Perspectives From Science and Technology Studies*. Edited by Flavia Padovani, Alan Richardson, and Thomas Uebel, 19–33. Cham: Springer, 2015.
11 Koskinen, Inkeri. "Objectivity in contexts: withholding epistemic judgement as a strategy for mitigating collective bias." *Synthese*, 199 (April, 2021): 211–225.
12 Cartwright, Nancy, et al. *The Tangle of Science: Reliability Beyond Method, Rigour, and Objectivity*. Oxford: Oxford University Press, 2022.

1 Tangled Understanding of Scientific Objectivity

The Proliferation of "Objectivity" in Science

The term objectivity is frequently used across science, policy, journalism, and everyday life as if it were a self-evident benchmark of truth or reliability. Yet a closer look reveals that even within scientific practice alone, "objectivity" is not a single, unified concept but a fluid and evolving constellation of ideals, spanning metaphysical claims about reality, procedural norms, epistemic virtues, and socially embedded conventions. Far from serving as a stable anchor, the notion has proliferated to such an extent that it now risks dissolving into conceptual ambiguity, and its coherence is stretched thin by competing and sometimes incompatible meanings. Consider just a few of the ways "objectivity" are used in philosophical and scientific contexts.

Objective world, reality, or properties: This sense of objectivity is exemplified by the title of Robert Nozick's book *Invariances: The Structure of the Objective World* (2001)[1]. It is sometimes referred to as physical objectivity. A well-known illustration comes from Albert Einstein, who wrote that the requirement of general covariance, a central postulate of the General Theory of Relativity, "takes away from space and time the last remnant of physical objectivity" (Einstein 1952, 117)[2]. In such contexts, the term "objective" is typically used to emphasize that the world, its objects, or their properties exist independently of human minds, perceptions, or conventional frameworks.

Objective description or representation: An objective description is one that faithfully corresponds to the facts or to the object it describes; likewise, an objective model accurately represents the physical world. This conception aligns with the correspondence theory of truth and with epistemic objectivism (Davidson 2001, 176–191)[3].

Objective mechanisms for acquiring knowledge: Objective mechanisms for knowledge acquisition are those insulated from personal biases, subjective preferences, or the influence of special interest groups. More broadly, results generated by standardized procedures that exclude or minimize subjective factors are often regarded as objective (see Daston and Galison 2007)[4].

DOI: 10.4324/9781003783787-2

Value-free or value-neutral descriptions: Descriptions that are value-free or value-neutral are often regarded as objective (see Douglas 2004)[5].

Intersubjective verifiability: The objectivity of a scientific theory is often understood to entail that its claims can be intersubjectively verifiable (see Popper 2002, 22; Quine 1992, 5)[6].

Reproducibility: Objective research results should be capable of being consistently replicated or verified across repeated trials (see "convergent sense of objectivity" in Douglas 2004)[7].

Socially situated objectivity: The objectivity of a scientific statement is not absolute but shaped by the norms, values, and perspectives of specific social groups, such as disciplinary communities, gender identities, or social classes. This view is reflected in Thomas Kuhn's account of paradigm-dependent science (Kuhn 1977)[8], social-political approaches to objectivity (e.g., Harding 2015; Longino 1990)[9], and feminist critiques that emphasize how situated knowledges influence what counts as objective (e.g., Lloyd 1995, 2005)[10].

Non-perspectival: Objective descriptions are non-perspectival (see Debs and Redhead 2007)[11].

The View From Nowhere: This concept, drawn from Thomas Nagel's *The View from Nowhere* (1986), describes an idealized method for attaining a more objective understanding of reality. Nagel emphasizes that "objectivity in whatever kind is not the test of reality. It is just one way of understanding reality" (Nagel 1986, 26). He observes that all human representations of the world are inherently perspectival—shaped by our specific location, biology, and cognitive makeup. However, objectivity increases as we step back from our immediate standpoint: when we situate a given perspective within a broader framework and articulate how it relates to that wider view, the latter becomes more objective. As Nagel puts it,

> A view or form of thought is more objective than another if it relies less on the specifics of the individual's makeup or position in the world, or on the character of the particular type of creature he is… Objectivity allows us to transcend our particular viewpoint.
>
> (1986, 5)

On this account, greater objectivity is not the elimination of perspective, but the conscious expansion and contextualization of it[12].

Mathematical objectivity: Mathematical knowledge, according to Gottlob Frege, is paradigmatically objective. He writes:

> Number is neither spatial and physical … nor yet subjective like ideas, but non-sensible, objective. Now objectivity cannot, of course, be based on any sense impressions, which as an affection of our mind is entirely subjective, but only so far as I can see, on *reason*.
>
> (Frege 1960, 37–38)

He goes even further, declaring, "There is nothing more objective than the laws of arithmetic" (Frege 1960, 115)[13]. For Frege, mathematical propositions are objectively true or false regardless of whether mathematical objects exist in the physical world or are entertained by any particular mind. Their truth is grounded not in empirical observation or psychological states, but in logical necessity—a view rooted in his broader logicist program, which holds that mathematics is reducible to logic (Reck 2007)[14].

The examples cited above represent only a few of the many ways the term objectivity is employed in scientific discourse, and they barely scratch the surface. Marianne Janack (2002)[15] identifies no fewer than 13 distinct senses of objectivity, including value neutrality, absence of bias, rationality, impersonality, factuality, disinterestedness, commensurability, and intersubjective agreement.

These varied notions of objectivity serve different functions. Some concern the correspondence between representations and reality; others focus on the relationship between a subject's cognitive states and their epistemic practices, such as the use of standardized mechanisms to produce reliable knowledge. Still others involve comparative judgments about cognitive outcomes (e.g., determining which of two claims is more objective), while yet others address normative epistemological questions about what objective knowledge ought to look like.

Moreover, these usages reveal that objectivity is invoked to characterize multiple dimensions of scientific cognition. At times, it describes the ontological status of the object under investigation, particularly its independence from the knowing subject. At other times, it pertains to the epistemic status of cognitive outputs, such as how well a theory corresponds to reality or how free it is from the idiosyncrasies of a particular perspective. In some contexts, objectivity refers to an evaluative stance, such as assessing a theory in a value-neutral manner. And in still others, it denotes features of the research process itself, such as the requirement that scientific inquiry employ procedures designed to minimize bias or subjective influence.

Clearly, the objects of cognition, the mechanisms of inquiry, the relationship between knowledge and the world, and attitudes toward cognitive results are all integral components of scientific practice. Scientific research is a complex whole composed of these interrelated aspects. Consequently, the various senses of objectivity are not isolated; they overlap and interact, although sometimes conflict.

For instance, the "objective mechanisms" invoked in one usage may refer precisely to research procedures that embody the value-detached or value-neutral ideals described elsewhere. Yet even knowledge produced through such supposedly neutral mechanisms can be contested. Different researchers may regard each other's conclusions as non-objective—not because of flaws in method but due to divergent interpretive frameworks. Consider quantum mechanics: scientists may observe the same phenomenon (e.g., electron double-slit interference) using identical, rigorously controlled methods, yet offer incompatible interpretations based on differing views of quantum theory.

Likewise, a claim may be intersubjectively verifiable because it arises from a shared, standardized procedure—but intersubjectively agreed data does not guarantee objectivity in a deeper sense. Even when subjects follow the same protocol, they do so from distinct epistemic standpoints. Their perspectives, shaped by training, theoretical commitments, or conceptual schemes, can lead to divergent understandings of the same data. Determining which perspective yields more objective knowledge is often far from straightforward. Thus, objective mechanisms do not automatically yield objective results.

In short, scientific discourse is saturated with entangled, context-dependent meanings of objectivity. This plurality poses a significant challenge to those who treat objectivity as a benchmark for scientific legitimacy. When a concept encompasses so many competing standards, it risks becoming vacuous: too many criteria may amount to no criterion at all.

Classifications of Various Notions of Objectivity

Confronted with the multitude of ways "objectivity" is used, a natural first step is to pursue a systematic classification of these varied usages. It is therefore unsurprising that researchers have long worked to group them into coherent categories and construct integrative frameworks, not merely to clarify what it means to describe something as "objective" but also to lay the groundwork for deeper philosophical reflection on the nature, scope, and limits of objectivity itself. What follows are several representative examples of such classificatory efforts.

Allan Megill's Four Senses of Objectivity

Allan Megill (1994)[16] offers a classification that distinguishes four interrelated yet distinct senses of objectivity, reflecting the multifaceted ways in which the term has been deployed across philosophical and scientific discourse.

At one extreme lies *the absolute sense of objectivity*, rooted in the modern epistemological ideal that knowledge should "representing things as they really are," unmediated by human perception, intention, or interpretation. In this view, subjectivity is not merely incidental but inherently distorting. Thus, any trace of personal bias or "good will" must be rigorously purged to achieve absolute objectivity. This sense of objectivity assumes a sharp dichotomy between subject and object, where truth emerges only when the former is entirely neutralized. Megill acknowledges that the absolute sense of objectivity, offering "a view from nowhere," is deeply paradoxical. The project of "representing things as they really are" rests on a tension between two inseparable yet conflicting dimensions: an ontological one—aiming to capture reality in its true, mind-independent form—and an epistemological one—since any access to that reality necessarily depends on representation itself. In other words, we cannot reach reality except through the very representational acts that, by their nature, mediate and potentially distort it. Although we may never actually attain such an absolutely objective view, as rational beings, we can be expected to converge toward an ever-closer approximation of it. Absolute objectivity, then, is not absolute in the sense of guaranteed certainty or infallibility, but rather in the normative grip it ought to exert on us—as an ideal that guides and constrains our reasoning precisely because we are rational.

In contrast to the stringent demands of absolute objectivity, *disciplinary objectivity* shifts the focus from universal criteria of validity to particular—yet still authoritative—standards internal to a given field. It does not presuppose the eventual convergence of all inquirers of good will; rather, it rests on the more immediate and practical consensus among accredited practitioners within a specific discipline. Disciplinary objectivity acknowledges that the complete elimination of human factors is neither feasible nor necessary. Instead, objectivity is secured through agreement among competent members of a well-defined scholarly community who share common methodological norms, theoretical commitments, and evaluative standards. On this view, objectivity is not a metaphysical guarantee grounded in correspondence to a mind-independent reality but a social achievement forged through collective judgment and disciplined practice. Disciplinary objectivity is deeply entangled with the dynamics of the modern academic enterprise, which is sharply segmented by discipline and field, and continually fractured by competing claims to authority. The system is inherently multicentered, with academic authority constantly contested, reconfigured, and reimposed. In the absence of a neutral or universally valid standpoint, disciplinary objectivity functions as a crucial form of epistemic legitimacy.

Unlike absolute and disciplinary objectivity which both regard subjectivity negatively, treating it as something to be minimized, neutralized, or excluded in the pursuit of objective knowledge, *dialectical objectivity* rejects the rigid dichotomy between subject and object altogether. It embraces subjectivity not as a contaminant but as an *indispensable* condition for the very constitution of objects of inquiry. Central to this perspective is a preference for doing over viewing: knowledge arises not through passive observation but through the dynamic, ongoing interaction between knower and known. On this account, objectivity is neither a static property of accurate representations nor a fixed standpoint from which the world is correctly perceived; rather, it is an emergent quality of relational processes, shaped by the active and transformative engagement of subjects with the world through research practices.

Finally, *procedural objectivity* functions less as a theory of truth, unlike absolute or disciplinary objectivity, and more as a regulative ideal. It demands strict adherence to transparent, non-manipulative methods, which are methods designed to ensure that data are not artificially molded to fit predetermined conclusions. At its core lies a commitment to methodological integrity: researchers must refrain from distorting evidence through selective reporting, biased instrumentation, or rhetorical sleight-of-hand. The governing metaphor of procedural objectivity is neither visual—like the "view from nowhere" associated with absolute objectivity—nor action-oriented, as in dialectical objectivity. Instead, it is tactile, though in a negative register: its implicit imperative is "hands off!" Its unofficial motto might well be "untouched by human hands"—a phrase that captures its vision of neutrality not as passive detachment but as disciplined restraint in the handling of evidence.

Megill emphasizes that the four senses of objectivity—absolute, disciplinary, dialectical, and procedural—are conceptual types that frequently intermingle in practice. Moreover, his classification is modest: it aims to help us navigate the current multidisciplinary discourse on objectivity, not to offer any definitive "resolution" to "the problem of objectivity."

Heather Douglas's Three Modes of Objectivity

Heather Douglas (2004)[17] argues that some traditional notions of objectivity, such as the idea of objective knowledge existing independently of mind, are not operationally viable. Instead, she focuses on operational articulations of objectivity and distinguishes three major modes, which together encompass eight distinct senses.

The first mode of objectivity concerns cognitive *interactions between humans and the world* that underpin scientific activity—whether in controlled experiments or everyday engagement. These interactions aim to "get at objects" and achieve understanding of real entities. It includes two senses: manipulable objectivity and convergent objectivity.

Manipulable sense of objectivity does not require that cognitive outcomes correspond to mind-independent reality. What matters is that specific interaction processes *reliably produce* related phenomena. Objectivity here is demonstrated by the capacity to intervene in nature predictably and effectively. A famous example comes from Ian Hacking's *Representing and Intervening*: once scientists can use electron microscopes to manipulate electrons and reliably produce consistent images, they no longer question the objectivity of electrons (Hacking 1983, 263)[18]. In short, manipulable objectivity prioritizes practice over representation.

Convergent sense of objectivity pertains to the outcomes of repeated cognitive activities. When a result consistently emerges *across diverse contexts* or through the use of a variety of methods, confidence in its reliability increases, leading observers to regard it as objective. The strength of such reliability claims depends crucially on the independence of the techniques employed, though one may still question the possibility of correlated errors or shared biases across methods.

The second mode of objectivity shifts focus from subject–world interactions to the role of values within an individual researcher's reasoning process. It asks: What does it mean to say that a person's judgment, conclusion, or review is "objective"? It comprises three senses: detached objectivity, value-free objectivity, and value-neutral objectivity.

Detached objectivity holds that personal biases or cognitive inflexibilities hinder the acquisition of accurate knowledge about independently existing entities. It therefore recommends maintaining a critical distance or detachment between the knower and the object of inquiry. Such detachment helps prevent researchers from desiring a particular outcome too strongly or fearing an unwelcome result so deeply that they become unable to perceive it clearly.

Value-free objectivity extends detached objectivity by asserting that all values are inherently subjective and that any intrusion of value judgments into the research process distorts findings, thereby undermining their objectivity. Consequently, it demands the complete exclusion of values or "bias" from scientific inquiry. Douglas regards this ideal as controversial. She argues that values can play multiple legitimate roles in scientific reasoning, beyond merely displacing evidence. Especially, non-epistemic values, such as social, ethical, or political considerations, can inform certain aspects of scientific practice, including decisions made during the internal stages of research. For example, when scientific inquiry leads to the development of a new technology, such values ought to be consulted to assess whether the technology is socially desirable or ethically justifiable. Likewise, non-epistemic values may legitimately constrain methodological choices, for instance, by imposing ethical limits on the use of human subjects in experimentation. Accordingly, Douglas contends that scientists need only remain value-free in the selection and interpretation of empirical evidence (Douglas 2000)[19].

Rather than demanding the absence of values, this sense requires researchers to adopt a balanced stance with respect to a spectrum of competing values. It is inherently reflective: the researcher must understand the range of plausible value judgments relevant to the issue and strive for impartiality or neutrality in their treatment. Objectivity here lies not in value elimination but in *fair* adjudication.

The third mode of objectivity examines how groups, such as scientific communities, achieve objective outcomes through social processes[20]. It moves beyond individual cognition to focus on collective validation and consensus. It includes three senses: procedural objectivity, concordant objectivity, and interactive objectivity.

Unlike Megill's conception of "procedural objectivity" which emphasizes strict adherence to transparent, non-manipulative methods, Douglas uses the term to denote the *uniformity in procedures*: if any competent researcher follows the same procedure, they should arrive at the same result. On this view, a finding achieves objectivity only when it can be replicated across different investigators. Concordant objectivity emphasizes agreement in outcomes rather than uniformity in method. It aligns with the common understanding of "inter-subjectivity": when a group of competent observers independently concur on a particular observation, that observation is regarded as objective. While procedural objectivity, defined as strict reproducibility through identical methods, can help achieve concordant objectivity, it is not necessary for it. In fact, a more robust form of concordant objectivity arises when different researchers, using diverse approaches or methodologies, nonetheless arrive at consistent conclusions. Such convergence across heterogeneous practices provides stronger epistemic warrant than mere replication under identical conditions. Nevertheless, Douglas cautions that even strong concordance does not guarantee truth: consensus can also result from collective illusions or shared biases.

Finally, rather than focusing on consensus *per se*, the interactive sense of objectivity highlights the dynamic, dialogical nature of scientific practice. It is not enough for participants simply to assent to an observation or result; instead, they are required to engage in sustained dialogue—debating, critiquing, sharing data, and working through disagreements to uncover their sources. On this view, scientific objectivity emerges not from static agreement but from an ongoing process of interaction, made possible by transparency, mutual accountability, and openness to scrutiny[21].

As Douglas points out, the eight senses of objectivity differ significantly in both meaning and context of applicability, making it implausible that any one sense can be reduced to another. For instance, the distinctions among detached, value-free, and value-neutral objectivity are robust enough to rule out mutual reducibility. Douglas argues that these various senses of "objectivity" collectively constitute a set of normatively significant criteria that we typically invoke when deciding whether to label a claim or practice as objective. It is precisely this plurality and complexity that account for both the term's flexibility in use and the strength of its normative force. The multiple senses of objectivity provide diverse grounds for trusting a claim and for justifying that trust to others.

The classifications proposed by Megill and Douglas are just examples among many similar attempts to map the diverse meanings of objectivity[22]. The purpose of presenting these classifications in this section is to highlight both the complexity of objectivity's entangled meanings and the difficulty of arriving at a unified or general conception of objectivity, even through systematic attempts to categorize its varied applications.

Galison and Daston's Historical Genealogy of Scientific Objectivity

Beyond the conceptual tangle of competing definitions, the very meaning of objectivity has shifted dramatically over time—a transformation meticulously documented by Lorraine Daston and Peter Galison in their 2007 work *Objectivity*[23].

Daston and Galison identify three historically contingent regimes of scientific representation, each anchored in distinct epistemic virtues: truth-to-nature, mechanical objectivity, and trained judgment. Crucially, these are not successive stages that cleanly replace one another; rather, they accumulate, overlap, and interact, much like "new stars winking into existence, not replacing old ones but changing the geography of the heavens" without extinguishing the old (Daston and Galison 2007, 18)[24].

Before the rise of modern objectivity in the mid-nineteenth century, scientific illustration was guided by what Daston and Galison term "truth-to-nature." Rather than striving for photographic fidelity to individual specimens, scientists aimed to depict idealized types that captured the essential character of a species or a phenomenon. An anatomist, for instance, might synthesize observations from dozens of frogs to produce a single "perfect" image that deliberately excluded anomalies to embody the universal form. In this paradigm, truth resided not in empirical particulars but in conceptual universals. Achieving this required more than mere observation: it demanded artistic skill, and cultivated intuition—qualities seen as essential for discerning nature's deeper order. Objectivity, in this context, entailed active idealization, not passive recording. As Daston and Galison note, "allegiance to truth-to-nature required that the naturalist be steeped in but not enslaved to nature as it appeared" (Daston and Galison 2007, 59)[25]. In practice, truth-to-nature encompassed a spectrum of interpretive *interventions*. Scientists sought "typical" phenomena that pointed back to an underlying *Typus* or "archetype," from which individual instances could be conceptually derived. Yet because any single observation could be misleading, the search for the typical necessitated systematic observation in series. The goal was never the raw datum itself, but the reasoned image that distilled it: whether conceived as the *Typus*, an ideal form, a characteristic exemplar, or a statistical average, truth-to-nature resided in the idea within the observation, not in the observation alone.

This ideal of truth-to-nature gave way in the mid-nineteenth century to mechanical objectivity, a new epistemic regime driven by the advent of imaging technologies, especially photography, and a deepening suspicion of human subjectivity. Where truth-to-nature had relied on a scientist's intuition to intervene, idealize, and distill nature's essence, mechanical objectivity demanded the opposite: rigorous resistance to all forms of human intervention. The new imperative was to let nature "speak for itself" through standardized, automatic instruments that minimized or ideally eliminated human mediation. Scientists were urged not to interpret, synthesize, or idealize but simply to record what the apparatus revealed. The cardinal virtue was now self-restraint: the disciplined refusal to impose meaning, pattern, or form onto raw data. To intervene, even in pursuit of the "typical" or the "essential," was now seen not as insight but as a dangerous lapse into subjectivity. Thus, while truth-to-nature saw intervention as necessary for revealing deeper truths, mechanical objectivity defined objectivity precisely by its rejection of intervention. Fidelity to empirical particulars, not abstraction toward ideals, became the hallmark of scientific integrity.

By the early twentieth century, a third epistemic regime emerged: trained judgment. As scientific fields grew increasingly complex, raw data alone proved inadequate. Instruments could generate images, but those images often required expert interpretation to become meaningful or usable. Objectivity was no longer equated with mechanical neutrality or idealized universals; instead, it came to reside in the reliability of expert consensus within specialized communities. What counted as "objective" was now determined less by the output of an instrument or the abstraction of an ideal form, and more by the soundness of professional judgment cultivated through years of training and experience. Mechanical fidelity was necessary but insufficient: complex arrays of visible phenomena needed to be smoothed, refined, classified, and contextualized before they could serve any scientific purpose. In place of the "many-eyed sight" of truth-to-nature (which sought universal types) or the "blind sight" of mechanical objectivity (which aimed to eliminate human agency), scientists now relied on what Daston and Galison describe as a kind of "physiognomic sight"—a cultivated capacity to recognize patterns, highlight salient features, and grasp relationships that could not be reduced to mechanical procedure (Daston and Galison 2007, 314)[26]. Accordingly, an increasing number of scientific inquiries began to invoke trained judgment explicitly, grounding their authority not in automation or idealization, but in the experiential knowledge and communal standards of expert practitioners.

The historical analysis by Galison and Daston reveals that objectivity is not a timeless epistemic ideal but a mutable virtue, continually reshaped in the history of science. The coexistence of multiple senses of objectivity today itself is a legacy of this layered history.

"Let's Not Talk About Objectivity"

The historical genealogy of "objectivity," alongside numerous attempts to classify its varied senses, reveals that the term is rich in meaning yet stubbornly resistant to definition. Taxonomic efforts have done much to map this conceptual terrain[27]. Yet for all their descriptive richness, they consistently fall short of answering the most basic question: *What is objectivity, especially scientific objectivity*?

As Ian Hacking (2012) observed in his review of Daston and Galison's book *Objectivity*:

> At the end of 415 pages of text, plus endless notes and gorgeous illustrations, we still are not told what objectivity is![28]

This remark captures a growing unease: even after exhaustive classifications, no unifying core emerges. Worse, the various senses of objectivity are not merely diverse. They are often mutually incompatible.

Consider the deep tensions that run through the literature. Some accounts, such as "mechanical objectivity" or "value-free science," assume that an individual researcher, armed with disciplined methods or epistemic virtue, might achieve objective knowledge in isolation. Others insist that objectivity is inherently social, emerging only through intersubjective validation among communities of practitioners. Similarly, while certain ideals demand the total exclusion of subjectivity, others recognize that interpretation, judgment, and even cultivated intuition are indispensable to scientific practice, suggesting that a refined form of subjectivity is not just compatible with objectivity but constitutive of it.

These are not marginal disagreements on the definition of objectivity or the irreducibility of different senses of objectivity. They reflect fundamental divergences about the nature of knowledge, the role of the scientist, and the very aims of science. Consequently, taxonomic projects may not simply fail to produce a unified concept but also may further demonstrate that no such concept is possible.

This insight leads some philosophers to abandon the search altogether. Hacking (2015), echoing Wittgenstein's notion of "family resemblance," argues that "objectivity" functions like "game," "art," or "democracy": a cluster concept held together not by a shared essence but by overlapping similarities and contextual affinities. Just as there is no single feature common to all games, there may be no essential property uniting all uses of "objectivity." Attempts to define it are thus misguided—not because we lack data but because the concept itself resists essentialist treatment. Hacking goes further. He suggests that when we ask: "Can we trust medical research when it is funded by pharmaceutical companies?"; and "whose research in climate science meets the standards of scientific objectivity?" what we really mean is: do we trust it? Hacking contends that invoking "objectivity" adds little to such debates. When we question the objectivity of climate science, we're usually worried about political interference; when we scrutinize industry-funded drug trials, our concern is corporate bias. In both cases, the real issue is trustworthiness—not an abstract metaphysical property. If that's so, why not speak directly about trust (or lack thereof) rather than invoke the ambiguous conception of "objectivity"? Thus, Hacking provocatively concludes: "Let's not talk about objectivity."[29] We can and should address concerns about bias, transparency, or methodological integrity directly, without recourse to an overloaded and ambiguous term.

This view gains further support from two additional considerations: disciplinary variation and historical change. First, standards of objectivity differ markedly across scientific fields. What counts as "objective" in particle physics—say, detector calibration and statistical significance—may bear little resemblance to the criteria used in epidemiology, evolutionary biology, or climate modeling (Megill 1994; Padovani, Richardson, and Tsou 2015)[30]. These differences stem from divergent methodologies, epistemic goals, and institutional norms, suggesting that objectivity is not a universal standard but a locally negotiated ideal. Second, as Daston and Galison (2007) compellingly show, the very meaning of objectivity has evolved over time from "truth-to-nature," to "mechanical objectivity," and toward "trained judgment." If objectivity is historically contingent, then any taxonomy based on current or past usage is inevitably provisional and likely to be overtaken by future scientific practices.

All this raises a pressing practical question: *Which conception of objectivity should guide scientific practice today?* Should we prioritize mechanical neutrality, intersubjective consensus, methodological rigor, or social accountability? And who decides—the philosopher, the scientist, the policymaker, or the public? In contested domains like behavioral genetics, environmental risk assessment, or public health, the choice of which "objectivity" to invoke carries real-world consequences for credibility, legitimacy, and democratic deliberation.

Given these conceptual, historical, and normative complexities, the project of classifying objectivity in order to isolate a single, correct definition appears not only difficult but also potentially futile. The term is too plural, too fluid, and too entangled in values to be pinned down by taxonomy alone.

"Saving" Objectivity by Pragmatic Approaches

Do the profound difficulties in understanding "objectivity" mean that we should, as Ian Hacking provocatively suggests, simply discard the concept altogether? Probably not. On the contrary, "objectivity" remains deeply embedded in both scientific self-understanding and public discourse, where it functions as a shorthand for key epistemic benchmarks such as reliability, fairness, impartiality, and epistemic authority.

Even if we were to abandon the term "objectivity" altogether, the underlying task would still persist: we would still need to explain and account for the diverse virtues and aspirations commonly associated with science, robustness, reproducibility, transparency, and more. Indeed, there is a deep-seated human impulse to unify these plural ideals under some coherent, if not singular, framework. Thus, while Hacking may counsel silence on the word "objectivity," the broader philosophical project to understand the norms and values that shape scientific practice remains vital. The challenge is not merely to catalog the term's varied uses but to clarify what we value in science, and why. Rather than seeking a single, timeless definition, we must learn to use "objectivity" critically: asking, in each context, what work the term is doing, whose interests it serves, and which *specific epistemic or social ideals it is invoked to protect*. This is exactly the pragmatic approaches that some scholars have sought to salvage a workable notion of scientific objectivity through notions of *trust* or *reliance*.

A notable example is Inkeri Koskinen's (2021)[31] "risk account of scientific objectivity." She proposes:

> When we call X objective, we say that we *rely* on X, and that others can safely do so too, because we believe that important epistemic risks arising from our imperfections as epistemic agents have been effectively mitigated or averted.

For Koskinen, this "risk account" aims to unify the diverse, context-sensitive senses of objectivity found in contemporary literature under a single, essentially descriptive framework. According to Koskinen, every recognized "applicable" sense of objectivity either identifies a specific epistemic risk stemming from human cognitive or social limitations (e.g., bias, overconfidence, groupthink) or describes a strategy for mitigating such risks. Crucially, she distinguishes these agent-centered risks from general epistemic challenges, such as the world's resistance to inquiry, which do not undermine objectivity *per se*. This clarification reveals a subtly moral dimension in our talk of objectivity: it responds not to the difficulty of knowing the world but to our own fallibility as knowers.

John Stephen (2021, 2–4)[32] proposes a related triad of constraints for any plausible account of scientific objectivity: it must concern representation, identify something genuinely valuable, and be practically viable. This framework suggests that talk of objectivity is primarily concerned with practices of representation—how we depict, model, or reason about the world. Moreover, objectivity is not merely a descriptive label; it is epistemically valuable. We have good reason to strive for objectivity in our epistemic practices and to place greater *trust* in those that are objective. Crucially, the concept of objectivity is meant to be *action-guiding*: it should serve as a yardstick for evaluating, improving, and sometimes reforming scientific practices. For this to be possible, objectivity must be viable, not in the sense that perfect objectivity is attainable, but in the sense that it is always possible to be more or less objective. Even if full objectivity remains an ideal, the notion must allow for gradations so that progress toward it can be meaningfully pursued and assessed.

More recently, Nancy Cartwright et al. (2022)[33] take another pragmatic stance. They acknowledge that "objectivity" is frequently overloaded, functioning less as a precise theoretical concept and more as a slogan that bundles together normative commitments central to scientific practice, such as value neutrality and the minimization of subjective judgment. In this respect, Cartwright et al. echo Ian Hacking's skepticism: rather than debating what "objectivity" means, we should address concrete questions directly. Is a given study trustworthy? Are its methods transparent? Are institutional incentives aligned with epistemic integrity? Such questions are best evaluated through close scrutiny of methods, incentives, and transparency, not by appealing to an abstract or idealized notion of objectivity. Their advice is unequivocal: stop calling your method "objective." Explain what actually makes it good. They further situate "objectivity" among what might be called Ballung concepts—dense, normatively charged terms like "democracy," "poverty," or "social exclusion." These concepts are simultaneously "world-guided" and "action-guiding": they describe features of reality, but specifically those we care about and are committed to addressing. This dual character underpins their central insight: to call something "objective" is not merely to describe it; it is to endorse it as worthy of reliance. Consequently, the duty to be objective, framed in terms of "objectivity-as-we-know-it," carries a substantial burden of responsibility, including an epistemic duty. Difficult as it may be, if you have a duty to be objective here, you must—better yet, figure out—what truly matters in these very specific circumstances. What is required is not an objectivity whose character is pre-set by universal criteria but an objectivity that must be *discovered* anew for each context. This paves the way for their notion of "objectivity-to-be-found." On this view, objectivity is not something we apply from a template; it is something we uncover through careful, situated judgment—responsive to the local interplay of epistemic needs, ethical constraints, and practical realities.

All these pragmatic approaches reaffirm that we should not discard "objectivity" entirely. Though imprecise, the term functions as a flag, signaling commitments to integrity, accountability, and epistemic responsibility. The task is not to define it once and for all, but to unpack what it signifies in each case, and ensure those signals are earned, not merely asserted.

However, these pragmatic approaches still fail to fully confront the central challenge posed by Hacking. If what we genuinely care about is trust, reliance, or responsibility, why not speak directly in those terms and treat them as our primary action-guiding concepts, rather than invoking the ambiguous and historically freighted notion of "objectivity"? This leads to another question: Are we using "objectivity" merely as a convenient umbrella term—a kind of conceptual cap—to bundle together diverse concerns about trust, reliance, responsibility, etc.? Or are we still engaging with objectivity as a distinct philosophical concept, one with deep historical and philosophical roots? In other words, is "objectivity" being repurposed as a rhetorical shorthand for values we already care about, or does it retain its own rich and contested history in philosophy that cannot be simply reduced to mere reliance or trust.

Moreover, if objectivity is now being reconceived in terms of trust or reliance, a crucial asymmetry emerges: while it may be relatively straightforward to derive notions of trustworthiness from traditional ideals of objectivity, say, by showing how objective methods warrant confidence, it is far less clear that the reverse holds. That is, does trust itself point back to objectivity? Can a framework build primarily on reliance, robustness, or risk mitigation genuinely recover the normative and epistemic aspirations that originally animated philosophical discussions of objectivity?

Historically, objectivity was not just about whether a claim works or whether we can depend on it; it was tied to deeper questions about how knowledge relates to the world, how inquiry transcends individual perspective, and how science achieves a kind of authority that goes beyond mere consensus or utility. To ground objectivity solely in practices of reliance risks flattening these ambitions into instrumental criteria (e.g., effectiveness, stability, or predictability) without addressing the original philosophical stakes: truth, justification, and the possibility of knowledge.

This tension raises a further, more pragmatic question: What are these contemporary projects on saving objectivity really trying to achieve? Are they attempts to preserve or rehabilitate the ideal of objectivity itself by refining and defending a venerable philosophical concept in light of new challenges? Or are they, more modestly (or perhaps more urgently), efforts to defend science as a social and epistemic enterprise in a time of escalating crisis in trust, marked by "misinformation," institutional erosion, and widespread public skepticism?

In other words, is "objectivity" being upheld for its own intellectual and normative sake, as a concept worth salvaging on philosophical grounds? Or has it become a strategic placeholder, a familiar banner under which to rally support for scientific integrity, even if its theoretical coherence is increasingly tenuous? If the latter, then the persistence of "objectivity" in discourse may say less about its conceptual viability and more about our collective need for a shared language of epistemic trust in uncertain times.

Notes

1 Nozick, Robert. *Invariance: The Structure of the Objective World*. Cambridge: Belnap Press of Harvard University Press, 2001.
2 Einstein, Albert. "The foundation of the General Theory of Relativity." In *The Principle of Relativity*. Edited by Hendrik Lorentz, et al., 111–164. New York: Dover, 1952.
3 Davidson, Donald. *Subjective, Intersubjective, Objective*. Oxford: Clarendon Press, 2001.
4 Daston, Lorraine and Peter Galison. *Objectivity*. Cambridge, MA: MIT Press, 2007.
5 Douglas, Heather. "The irreducible complexity of objectivity." *Synthese*, 138 (February, 2004): 453–473.
6 Popper, Karl. *The Logic of Scientific Discovery*. London: Routledge, 2002; Quine, Willard Van. *The Pursuit of Truth*. Cambridge: Harvard University Press, 1992.
7 Douglas, Heather. "The irreducible complexity of objectivity." *Synthese*, 138 (February, 2004): 453–473.
8 Kuhn, Thomas. "Objectivity, value judgment, and theory choice." In *The Essential Tension – Selected Studies in Scientific Tradition and Change*. Edited by Thomas Kuhn, 320–339. Chicago: The University of Chicago Press, 1977.
9 Harding, Sandra. *Objectivity and Diversity*. Chicago: University of Chicago Press, 2015. Longino, Helen. *Science as Social Knowledge*. Princeton, NJ: Princeton University Press, 1990.

10 Lloyd, Elisabeth. "Objectivity and the double standard for feminist epistemologies." *Synthese*, 104 (September 1995): 351–381; Lloyd, Elisabeth. *The Case of the Female Orgasm: Bias in the Science of Evolution*. Cambridge: Harvard University Press, 2005.

11 Debs, Talal and Michael Redhead. *Objectivity, Invariance, and Convention: Symmetry in Physical Science*. Cambridge: Harvard University Press, 2007.

12 Nagel, Thomas. *The View From Nowhere*. New York: Oxford University Press, 1986.

13 Frege, Gottlob. *The Foundations of Arithmetic: A Logico-Mathematical Enquiry into the Concept of Number.* New York: Harper Brothers, 1960.

14 Reck, Erich. "Freg on truth, judgment, and objectivity." *Grazer Philosophische Studien*, 75 (January 2007): 149–173.

15 Janack, Marianne. "Dilemmas of objectivity." *Social Epistemology*, 16 (November, 2002): 267–281.

16 Megill, Allan. "Four senses of objectivity." In *Rethinking Objectivity*. Edited by Allan Megill, 1–29. Durham: Duke University Press, 1994.

17 Douglas, Heather. "The irreducible complexity of objectivity." *Synthese*, 138 (February, 2004): 453–473.

18 Hacking, Ian. *Representing and Intervening: Introductory Topics in the Philosophy of Natural Science*. New York: Cambridge University Press, 1983.

19 Douglas, Heather. "Inductive risk and values in science." *Philosophy of Science,* 67 (December 2000): 559–579.

20 For further discussions on social characters of science and their relationship to objectivity, see Kitcher, Philip. *The Advancement of Science: Science Without Legend and Objectivity Without Illusions.* New York: Oxford University Press, 1993; Longino, Helen. *Science as Social Knowledge: Values and Objectivity in Scientific Inquiry.* Princeton, NJ: Princeton University Press, 1990.

21 Compared with the other seven senses of objectivity identified by Douglas, the interactive sense is more like a framework or a set of procedural norms demanded in the name of objectivity, such as data sharing, open discussion of theories, public accessibility of models for scrutiny, and, where possible, experimental replication.

22 For other classifications of objectivity, see: Mühlholzer, Felix. "On objectivity." *Erkenntnis*, 28 (1988): 185–230. Agazzi, Evandro. *Scientific Objectivity and Its Contexts.* Cham: Springer, 2014; Reiss, Julian, and Jan Sprenger. "Scientific objectivity." In *The Stanford Encyclopedia of Philosophy* (Winter 2020 Edition). Edited by Edward N. Zalta; Tsou, Jonathan, Alan Richardson, and Flavia Padovani. "Introduction: objectivity in science." In *Objectivity in Science: New Perspectives from Science and Technology Studies*. Edited by Flavia Padovani, Alan Richardson, and Jonathan Tsou, 1–15. Cham: Springer International Publishing, 2015; Axtell, Guy. *Objectivity*. Cambridge: Polity Press, 2016.

23 Daston, Lorraine and Peter Galison. *Objectivity*. Cambridge, MA: MIT Press, 2007.

24 Daston, Lorraine and Peter Galison. *Objectivity*. Cambridge, MA: MIT Press, 2007.

25 Daston, Lorraine and Peter Galison. *Objectivity*. Cambridge, MA: MIT Press, 2007.

26 Daston, Lorraine and Peter Galison. *Objectivity*. Cambridge, MA: MIT Press, 2007.

27 On the contrary, the more thoroughly we attempt to classify the various uses of "objectivity," the more elusive its definition becomes.

28 Dear, Peter, et al. "Objectivity in historical perspective." *Metascience*, 21 (January 2012): 11–39.

29 Hacking, Ian. "Let's not talk about objectivity." In *Objectivity in Science: New Perspectives From Science and Technology Studies*. Edited by Flavia Padovani,

Alan Richardson, and Jonathan Tsou, 19–33. Cham: Springer, 2015. For "family resemblance," see Ludwig Wittgenstein, *Philosophical Investigations* (4th ed). Translated by Gertrude Anscombe, Peter Hacker, and Joachim Schulte. Oxford: Wiley-Blackwell, 2009, §§66–68.

30 Padovani, Flavia, Alan Richardson, and Jonathan Tsou (Eds). *Objectivity in Science: New Perspectives from Science and Technology Studies*. Cham: Springer International Publishing, 2015.

31 Koskinen, Inkeri. "Objectivity in contexts: withholding epistemic judgement as a strategy for mitigating collective bias." *Synthese*, 199 (April 2021): 211–225.

32 Stephen, John. *Objectivity in Science*. Cambridge: Cambridge University Press, 2021.

33 Cartwright, Nancy, et al. *The Tangle of Science: Reliability Beyond Method, Rigour, and Objectivity*. Oxford: Oxford University Press, 2022.

2 Publicity as a Necessary Condition for Scientific Objectivity

A Hypothesis for a Necessary Condition of Scientific Objectivity

As the historical and taxonomic approach to the conception of scientific objectivity shows, constructing a single, universal conception of objectivity that subsumes and explains all its diverse uses appears challenging and even deeply problematic. The term "objectivity" carries a multiplicity of meanings shaped by shifting scientific practices, theoretical commitments, and cultural contexts. These conceptions often pull in different directions—epistemic, methodological, social—embedding conflicting demands that resist harmonization under one overarching definition.

In response to this impasse, I propose a "backward" approach to identifying what might be universal across most, if not all, plausible notions of scientific objectivity. Traditional strategies, as surveyed in the previous chapter, typically adopt a taxonomic or *inductive* method: they begin by cataloging the varied uses of "objectivity" in scientific discourse and everyday language, then attempt to sort these different notions into distinct types, sometimes culminating in an effort to distill a general or essential core from this diversity. Rather than beginning with the plurality of existing notions of objectivity, I start from *a conjectural hypothesis*: that there exists *a minimal, necessary condition* underlying any legitimate claim to scientific objectivity. This condition is identified through reflective inquiry into the indispensable preconditions of scientific knowledge itself. It is in this sense, by seeking the conditions that must be in place for objective knowledge to be possible, that I characterize my approach as transcendental.

Moreover, my hypothesis is not derived inductively from observed uses of the term "objectivity." Instead, it is advanced as a philosophical proposal about the constitutive structure of objective knowledge itself. Accordingly, I do not begin by classifying the various senses of objectivity. Rather, I directly propose a minimal, or necessary, condition as a conjecture. The task then becomes one of critical testing: Does this proposed necessary condition hold across the major conceptions of scientific objectivity?

DOI: 10.4324/9781003783787-3

In spirit, this method echoes Karl Popper's model of scientific inquiry in *Conjectures and Refutations*[1]. Like Popper, I advance a bold conjecture and submit it to rigorous scrutiny. Of course, there is a crucial difference: Popper's framework operates mainly within the natural sciences, where hypotheses are tested against empirical observations. My project fundamentally concerns the conceptual architecture of scientific objectivity itself. Consequently, the "evidence" against which the hypothesis is evaluated consists not of experimental data but of conceptual coherence and compatibility with major senses of scientific objectivity.

By adopting this hypothetical, rather than inductive-taxonomic, approach, I aim to circumvent the stalemates of classification of various notions of objectivity and move toward a robust understanding of what objectivity, at its core, requires. To be more specific, my *hypothesis* is: *publicity is a necessary condition for scientific objectivity.*

Publicity condition: For any scientific claim to qualify as objective, it must be, in principle, accessible to multiple individuals. It must reflect an understanding of a *same* subject matter that is not confined to a single subject's private experience.

This condition does not demand that objective knowledge be actually universally agreed. Rather, it insists that objectivity is impossible without *public access*. A claim that cannot, even in principle, be scrutinized or evaluated by others because its basis is "locked" within a solitary consciousness would fail to meet the minimal threshold of scientific objectivity. We would and should be deeply skeptical of any such claim, whether in the form of method, data, or representation. Indeed, if objectivity could rest on private intuition, the very *normative* force of the concept would collapse.

Taking publicity as a necessary condition for scientific cognition seems evident for two complementary lines of reasoning.

First, science is inherently a public and collective enterprise. From its institutional structures (peer review, collaborative labs, etc.) to its epistemic norms (reproducibility, transparency, communal validation), scientific practice is designed to transcend individual limitations. As such, its cognitive products, scientific knowledge, are expected to be public: intelligible and accessible by multiple or even any competent inquirers, regardless of origin. In a domain defined by cooperation across time, space, and culture, it is natural and necessary that scientific knowledge is not the private possession of any one mind.

Moreover, if scientific knowledge aims to represent a reality that is mind-independent, then those representations must themselves, in principle, be capable of being verified by multiple observers. A "representation of reality" that only one person can access or understand ceases to function as a representation of an objective world as it becomes indistinguishable from a personal hallucination or subjective impression. Thus, requiring the publicity condition is not merely a pragmatic ideal; it seems also a constitutive requirement for representing *a shared, external world.*

Second, the conceptual opposition between objectivity and subjectivity also seems to further support this conjecture. Subjectivity is often associated with private aspects: it relates to states, judgments, or experiences that are internal, perspectival, and potentially inaccessible to others, entailing a belief in the autonomy of individuals that is free from others, and are shaped by personal interests, emotions, or unique biographical circumstances. The antonym of "private" is "public." If objectivity is to serve as the counterpoint to subjectivity, then it is expected to involve a move beyond the private self toward a standpoint that is, at least in principle, available to multiple subjects. Publicity, therefore, might not be an accidental feature of objectivity but a structural one.

However, to be clear: publicity is proposed here as a necessary, but *not* a sufficient, condition for scientific objectivity. Public accessibility—multi-subjective or even widespread acceptance—alone does not guarantee the kind of robust objectivity demanded by many discourses in science or philosophy. After all, many widely disseminated claims are demonstrably false, deeply biased, or even morally pernicious, for instance, some of the information on social media. Such content may be publicly accessible, widely shared, and even collectively endorsed by large communities, yet it remains far from objective, say, in the sense of value-free. It may reflect groupthink, ideological conformity, or coordinated "disinformation" rather than epistemic reliability. Also, the history of science abounds with examples of collectively endorsed theories later overturned as false: phlogiston chemistry, geocentric cosmology, or even modern cases of fraudulent research that temporarily gained broad credibility. More insidiously, scientific communities can fall prey to shared biases, institutional inertia, or collective blind spots—situations in which consensus masks error rather than corrects it. These cases underscore a crucial lesson: publicity must be coupled with other epistemic virtues to achieve scientific objectivity in some sense. Publicity opens the door to some notions of scientific objectivity, and it does not substitute for them.

Nevertheless, while publicity alone cannot ensure objectivity in the sense of some notions of objectivity, it is a prerequisite for achieving objective claims. Without it, the very notion of scientific objectivity collapses into idiosyncrasy or solipsism. My claim is therefore narrower and simpler: *nothing can count as scientifically objective unless it is, in principle, public.*

Thus, although full-fledged scientific objectivity likely requires additional conditions, such as empirical adequacy, methodological rigor, logical coherence, and resistance to bias depending on in what exact sense the notion of scientific objectivity is used, I argue that publicity is the minimal threshold that must be crossed before any claim can even aspire to objectivity. Any knowledge that abandons publicity condition ceases to function as objective knowledge.

The Conception of "Publicity"

Before testing whether publicity condition can serve as a necessary or minimal condition across the major conceptions of scientific objectivity, it is essential to further clarify what I mean by "publicity." The term also carries various different meanings in everyday language and, especially, in social–political discourses, and conflating these with the sense relevant to scientific objectivity would risk confusion.

In traditional discussions of the public–private dichotomy, the term "private" indicates the realm of familial, the personal or intimate relations, while "public" signals the civic realm or realm of community outside of this personal one (e.g., the Habermasian "public sphere")[2]. This distinction dichotomy could date back to Aristotle's division between *polis* and *oikos*: the *polis* representing the public sphere of political life and collective deliberation, and the *oikos* designating the private, domestic sphere of the household and family[3].

In some recent social–political contexts, privacy sometimes is understood as a form of control over personal information, specifically, the right to determine *who has access to* aspects of one's identity, life, or data: "privacy provides one aspect of reputational control by allowing us to decide who we share various aspects of ourselves with," and "claims to privacy are claims to prevent certain information from being publicly available beyond one's audience of choice" (Hannah 2023)[4]. In this framework, "publicly available" simply means that a piece of information, say, a text, image, or data record, exists in a format accessible to many individuals, *regardless of how they interpret it or whether it is true*. Privacy-publicity in this sense is related to debates about whether, for instance, corporations should be permitted to harvest users' social media posts for targeted advertising, even if those posts were voluntarily made public (Rumbold and Wilson 2019)[5]. Crucially, in such discussions, privacy is tied to *ownership and consent*: it concerns who has the right to disclose or restrict access to information about oneself, not whether the information itself is epistemically reliable, intersubjectively verifiable, or reality grounded. A false rumor about someone may be "private" if shared only within a closed group, or "public" if posted online, but its truth value or epistemic status is orthogonal to its classification as private or public in this sense.

However, in the context of publicity as a necessary condition for scientific objectivity, "publicity" does not refer merely to the wide dissemination or formal accessibility of knowledge, such as that when a scientific textbook sits on a library shelf, it is physically or digitally available to anyone who opens it. Nor does it concern permissions, authorship rights, or data governance protocols. Instead, publicity here denotes an epistemic condition: *for a claim in science to be genuinely public, it must transcend the private cognition of any single individual*. This means it must be collectively accessible—capable of being recognized, assessed, and *endorsed by multiple competent inquirers*. Public knowledge, in this sense, is not defined by its availability but by its *multi-personal warrant*.

Therefore, by "publicity" as a necessary condition for scientific objectivity, I mean that scientific claims must *be formulated in such a way* that they are collectively accessible and capable of being recognized, and *potentially* endorsed by multiple competent inquirers. By its very nature, scientific knowledge cannot remain tethered to the mind or consciousness of its original author. A scientific claim must not merely reflect what one individual knows, believes, intuits, or experiences, even if that individual is a figure of exceptional insight like Albert Einstein. Also, no matter how mathematically elegant or intuitively compelling a theory may be, it does not enter the domain of scientific knowledge unless it is articulated in a form that enables uptake, scrutiny, and validation by others. Only through such communicability and multi-personal validity or endorsement does a private insight become public knowledge.

This is not merely a sociological observation about how science happens to be practiced in scientific communities. It is a constitutive feature of scientific objectivity itself. Objectivity requires that knowledge be depersonalized in its epistemic presentation, not in the sense of erasing credit or historical authorship, but in the deeper sense that its validity must not depend on the authority, identity, psychological state, or private experience of the person who first articulated it.

One further point merits emphasis. The publicity condition should not be conflated with universal consensus. To satisfy this condition, a scientific representation need not be accepted by all or most members of the scientific community. Rather, the publicity condition suffices that the claim be accessible to, and evaluable by, *multiple competent individuals beyond its originator*. In other words, *the minimal requirement is that the representation transcend the confines of a single subjectivity*. Of course, if a claim eventually achieves broad or even universal consensus among competent inquirers, that would constitute a significant epistemic achievement and certainly fulfill the publicity condition. But such a consensus is not required at the outset. The essential first step toward objectivity is simply that the claim be *detachable from its author*, formulated in a way that allows others to engage with, test, and potentially corroborate or refute it. This modest threshold, moving beyond the solitary "I," is about overcoming the limitations of privacy without demanding excessive or premature epistemic closure.

Publicity and Tangled Senses of Scientific Objectivity

Publicity as a minimal conception of objectivity offers an alternative to both the projects of exhaustive classification of various senses of objectivity and the skepticism that would abandon the notion of objectivity altogether. As argued earlier, when people view science as inherently a public and collective enterprise and also situate this view within the conceptual frameworks of the subjective/objective and private/public dichotomies, it seems plausible to propose publicity as a necessary condition for scientific objectivity. In what follows, I will examine whether publicity can indeed serve as a necessary condition across several major conceptions of scientific objectivity.

Objectivity as Intersubjective Agreement

Scientific objectivity is sometimes interpreted as intersubjective agreement on claims, models, or representations of the natural world. Yet the notion of intersubjectivity itself is far from monolithic; it is historically layered. As Edmund Husserl first articulated in his *Cartesian Meditations* (1991)[6], intersubjectivity originates not in consensus about facts or reality but in the phenomenological problem of other minds: how can one conscious subject recognize and relate to other conscious subjects (Crossley 1996, 1–7)[7]. For Husserl, while we have immediate, first-person access only to our own consciousness, we come to acknowledge others through an act of empathy, an imaginative analogical transfer whereby we project our own lived experience onto the perceived behavior of others. Empathy, in this view, is not mere emotional resonance but a cognitive synthesis of perception and imagination (Crossley 1996, 6)[8].

This foundational phenomenological concern later evolved into broader social and epistemological conceptions. Contemporary theorists offer divergent definitions: Wendelin Reich (2010), for instance, defines intersubjectivity as "a situation in which two or more persons share knowledge reflexively—that is, all know X and know that all others know this too"[9]. Paul Munroe (2019) describes it as "a shared perception of reality between two or more individuals… [and] a common understanding of what is going on."[10] Such accounts presuppose a basic epistemic constraint: human beings access reality only through their own senses, sight, hearing, touch, etc., however augmented by instruments like telescopes, microscopes, or particle detectors. Consequently, each individual's grasp of reality remains, at its root, subjective. And yet, most people also recognize that reality is not malleable by private will: no matter how fervently one insists that "blue" is "yellow," the external world resists such subjective redefinition. This tension, between the inescapability of subjective access and the stubborn independence of reality, poses a persistent challenge for theories of knowledge. Neither pure objectivity (as detached, "God's-eye" truth) nor radical subjectivity (as solipsistic experience) suffices to account for how we navigate and coordinate our understanding of the world. Intersubjectivity, however, emerges as a mediating epistemic stance[11], offering a socially grounded alternative: the best we can achieve is a mutually intelligible, communicable, and revisable understanding of shared phenomena.

In the philosophy of science, this idea takes a distinctive form: intersubjectivity is closely tied to testability. Karl Popper famously argued that "the objectivity of scientific statements lies in the fact that they can be intersubjectively tested" (Popper 2002, 22)[12]. For Popper, scientific knowledge is never fully verifiable, but its objectivity derives from its reproducibility, or being repeatably tested. As he writes:

> We do not take even our own observations quite seriously, or accept them as scientific observations, until we have repeated and tested them. Only by such repetitions can we convince ourselves that we are not dealing with a mere isolated "coincidence," but with events which, on account of their regularity and reproducibility, are in principle inter-subjectively testable.
>
> (Popper 2002, 23)[13]

This emphasis on repeatability shifts the locus of objectivity from metaphysical correspondence to epistemic accessibility: a claim is objective not because it mirrors reality in some absolute sense but because it can be scrutinized, replicated, and *challenged by all subjects*.

Given these notions of intersubjectivity, we may now ask: Does intersubjective objectivity require publicity as a necessary or minimal condition? The answer appears unequivocal. By definition, intersubjective knowledge demands that a claim be accessible to, assessable by, and recognizable among multiple subjects. Whether framed as Reich's reflexive mutual awareness ("all know X and know that all others know X") or Popper's reproducible observation, the very possibility of intersubjectivity presupposes that the relevant content, be it data, method, or inference, is not confined to a single mind. It must be expressible, communicable, and open to multi-personal verification. In short, it must be public.

Publicity here does not entail universal consensus or institutional endorsement; rather, it signifies minimal epistemic openness: the claim must be formulated in such a way that others can, in principle, engage with it. Without this condition—without the possibility of access and assessment by others—the notion of intersubjective agreement collapses into incoherence. One cannot agree, disagree, or even meaningfully discuss what remains forever locked within a private horizon.

Thus, for conceptions of scientific objectivity grounded in intersubjectivity, whether phenomenological, sociological, or Popperian, publicity functions not merely as a helpful ideal but as a constitutive prerequisite. It is the threshold that separates private insight from public knowledge, subjective conviction from scientifically admissible claim. To be intersubjective is, by necessity, to be public.

Procedural/Mechanical Objectivity

Let us now turn to the family of conceptions of scientific objectivity that centers not on the neutrality of the knowing subject but on the mechanical procedures of scientific cognition, including using mechanical instruments for measurements, standardized experimental protocols, modeling frameworks, statistical methods, formalized inference rules, etc. On this view, a scientific representation is deemed objective not because it emanates from a disinterested or "view-from-nowhere" mind but because it is generated through methodologically disciplined practices that are designed to be robust and standardized.

The core insight here is that objectivity is not achieved by eliminating human judgment altogether—indeed, human judgment remains indispensable at nearly every stage of scientific work—but rather by institutionalizing practices that systematically constrain, channel, and correct for *personal variability*. Such practices function as epistemic scaffolds: they do not remove the scientist from the process but embed their agency within a *norm-governed framework* that limits arbitrariness and enhances reliability. Classic examples include double-blind clinical trials, which prevent both researchers and participants from influencing outcomes through expectation bias; standardized calibration protocols for instruments like atomic clocks or mass spectrometers; and peer review, which subjects claims to communal evaluation.

As Allan Megill (1994) observes, this strand of objectivity emphasizes what he calls the "impersonality of procedure," often abstracting from the traditional epistemic aim of truth itself and focusing instead on adherence to rule-bound conduct[14]. In this sense, objectivity requires that scientists act not according to personal intuition or private conviction but in accordance with *publicly codified rules* that are, in principle, identical for all practitioners. Thus, while individual scientists may harbor diverse theoretical commitments, cultural backgrounds, or cognitive styles, the procedures they follow must be *uniformly applicable* and *interpretable across the community*. This reliance on standardized procedures can leave two possible distinct imprints on scientific cognition and representation.

First, such procedures may impose a *common* structural framework on how phenomena are perceived, categorized, and modeled. Much like Thomas Kuhn's notion of a paradigm, a shared methodological regime can shape what counts as a legitimate problem, a valid observation, or an acceptable explanation. For instance, the adoption of randomized controlled trials as the gold standard in medical research doesn't just test hypotheses, but it actively constructs a specific kind of evidence, marginalizing forms of knowledge that fall outside its procedural logic. In this way, standardized methods help constitute the objects and relations that populate scientific worlds.

Second, mechanical procedures may function as *epistemic filters*: only those aspects of the world that are detectable, quantifiable, and compatible with the operational logic of the instrument or protocol get registered as data. The procedures do not just process reality "neutrally"; they selectively transduce it, admitting only signals that resonate with their built-in sensitivities and thresholds. What lies outside this resonance, however real, remains epistemically invisible.

Crucially, for these methodological norms to serve as stabilizing, corrective, and legitimating mechanisms, they themselves must be public. That is, they must be collectively accessible, shareable, and open to critical scrutiny by multiple competent agents. A calibration protocol known only to its inventor, a statistical model whose code is proprietary and undocumented, or an experimental design whose criteria for inclusion are opaque, cannot function as genuine norms. At best, they are private prescriptions. These mechanical procedures serve as "*public standards*." The point of requiring standards that are public is that, by explicitly or implicitly professing adherence to those standards, the evaluation of the goals of inquiry is not ascertained privately but is open to communal interrogation. This evaluation may be performed by anyone, not just by members of the community sharing all standards (Longino 2002, 130–131)[15].

Thus, the very idea of mechanical procedures which transcends individual subjectivity presupposes publicity. Only when procedures are publicly articulated can they be replicated by multiple researchers, open to communal interrogation, etc. Within this procedural conception of objectivity, where instruments, algorithms, or rulebooks rather than minds provide a common ground or build *a common imprint*, a public sphere, into experience. Publicity therefore is a necessary condition for notions of objectivity in the procedural or mechanical sense. The objectivity of a scientific *representation* in certain senses might not be guaranteed merely by its production through a rule-bound system; rather, the system itself can only aspire to objectivity, if its rules, operations, and outputs are publicly accessible and multi-subjectively intelligible, and the public standards impose and "filter" a common imprint on results. In this sense, publicity is not an add-on to methodological rigor; it is its very foundation.

Perspective-Distanced Objectivity

A family of conceptions of scientific objectivity centers on the distancing or neutralization of perspective, whether individual or collective, in the processes of inquiry, representation, and evaluation. For these conceptions, objectivity is not defined by correspondence to a mind-independent reality alone but by the extent to which knowledge claims are freed from the distorting influence of situated perspective. These perspectives may be:

Individual or private: shaped by personal interests, cognitive biases, emotional dispositions, idiosyncratic values, or autobiographical contingencies;
Communal-but-particular: shared within specific groups such as scientific communities in Kuhn's sense, cultural traditions, institutional affiliations (e.g., corporate or state interests), gendered experiences, or disciplinary communities.

While communal perspectives are shared among members of a particular group—be it a scientific discipline, cultural tradition, institutional body, or social identity—they are not universal. Rather, they embody the norms, assumptions, values, and priorities of that specific collectivity. As such, they can inadvertently obscure, distort, or marginalize alternative ways of seeing the world.

Among conceptions of objectivity that aim to overcome communal perspectives, it is necessary first to overcome individual perspectives and achieve impersonality. After all, communal perspectives are aggregates or institutionalizations of individual perspectives. One cannot meaningfully transcend the biases of a group without first disciplining the subjective or private lenses through which its members perceive and interpret the world.

That said, the converse does not hold: successfully neutralizing individual perspective does not automatically eliminate communal bias. Scientists may scrupulously avoid personal pronouns, follow standardized protocols, and report findings in rigorously impersonal language, yet still operate within a paradigm that systematically excludes certain phenomena, questions, or forms of evidence.

Anyway, since overcoming individual perspective or impersonality is a prerequisite for overcoming communal perspective, it follows that publicity is also a necessary condition for any conception of objectivity that seeks to transcend group-bound assumptions. Therefore, across both individual and communal registers, publicity remains a necessary condition for objectivity.

Objectivity Centering on Subject Matter

Beyond the conceptions of objectivity centered on models of representation which are about how the subject matter is represented, such as intersubjectivity, impersonality, or communal procedures, there are also conceptions of objectivity that focus primarily on the subject matter[16]. I now examine whether the condition of publicity holds as a necessary condition for objectivity in this ontologically oriented sense.

One prominent such conception is mind-independence (or mental independence) or what is real. In this sense, a phenomenon is objective if its existence and nature are independent of minds, beliefs, or intentional states. According to this view, the basic constituents of the natural physical world, planets, oceans, trees, and bacteria are paradigmatically objective, qua mind-independent entities (setting aside philosophical idealism). By contrast, entities that are inherently mental or intentional, such as individual minds, beliefs, emotions, as well as social constructs like organizations, nations, languages, and scientific theories, are not constitutively mind-independent. Their existence and character depend on cognitive, linguistic, or social practices. Hence, they are not "objective" in this strict sense of mind-independence.

Now, does the condition of publicity apply to notions of objectivity in this context? At first glance, it appears that publicity is still a necessary condition for objectivity in this sense of mind-independence. After all, the very claim that planets, water, or trees are objective typically implies that they are collectively accessible, shareable in experience, and capable of being recognized and verified by multiple observers; otherwise, they might be hallucinations. Their mind-independence is not merely a metaphysical assertion but is epistemically supported by their public accessibility. However, a crucial distinction arises: while mind-independent entities are public, not all public entities are mind-independent. Scientific models as mathematical structures and symbols in textbooks, for instance, are paradigmatically public. They are communicated, taught, debated, and endorsed across scientific communities. Yet they are mind-dependent: they exist as products of human cognition, conceptual frameworks, and linguistic systems. Their truth or validity may be objective in other senses (e.g., intersubjective or procedural), but their ontological status is not independent of minds.

This shows that publicity is necessary but not sufficient for objectivity in the sense of mind-independence. Public accessibility is required to verify or attribute mind-independence (since private, unshareable experiences cannot ground claims about a mind-independent world), but publicity alone does not constitute mind-independence.

Moreover, this highlights an important point about the use of the term scientific objectivity in practice. While people acknowledge that scientific theories are human constructs that were initially developed by individual scientists and thus mind-dependent, they appeal to epistemic or methodological senses, such as reproducibility and multi-subjectively accessibility, all of which presuppose publicity as a foundational condition.

Publicity and the Impossibility of Private Objectivity

In previous discussions, publicity has been defined as an understanding of the subject matter that is not confined to a single subject's private experience, but is instead collectively accessible, shareable, and capable of being recognized and endorsed by multiple competent observers. This notion has turned out to be a necessary condition, if not always sufficient, for all major conceptions of scientific objectivity: whether framed in terms of intersubjectivity, perspective-detachment, mechanical reproducibility, or mind-independence. Across these diverse senses of objectivity, the requirement that knowledge be publicly available consistently resurfaces as a foundational epistemic threshold.

That said, it is crucial to emphasize again: publicity does not guarantee objectivity. Rather, publicity functions as the minimal precondition: without it, the very ascription of "objectivity" becomes meaningless. My aim is not to defend any full-fledged conception of objectivity or to argue that various conceptions of objectivity can be universally deduced from publicity, but to argue that all such conceptions of objectivity implicitly presuppose publicity as a baseline epistemic norm. Whether one prioritizes methodological rigor, representational neutrality, or ontological independence, the expectation remains constant: scientific knowledge or the subject matter must be accessible to, and assessable by, a community of inquirers.

Once we accept publicity as necessary for scientific objectivity, a pressing question arises:

> Can a single or solitary individual, working in isolation, achieve objective understanding all by herself?

At first glance, this seems plausible. After all, many scientific discoveries begin with a lone thinker. But upon closer inspection, even these "solitary" insights rely on publicly constituted frameworks. To unpack this, we must distinguish two subtly different sub-questions:

Q1: Can meaningful understanding be private?
Q2: Can an individual who already possesses public concepts achieve objective knowledge alone?

The first question leads us to Ludwig Wittgenstein's famous critique of the "private language" in *Philosophical Investigations* (§§243–271). By "private language," Wittgenstein means that "The words of this language are to refer to what only the speaker can know—to his immediate private sensations. So another person cannot understand the language" (Wittgenstein 2009, §243)[17]. Imagine a language whose words refer exclusively to the speaker's immediate private sensations, say, a symbol "S" that stands for a particular toothache only she can feel. Could such a language be meaningful?

Wittgenstein argues no. For a word to have meaning, there must be public criteria for its *correct use*—standards that allow others (and even the speaker at a later time) to distinguish between using the term correctly and merely thinking one is using it correctly. In the absence of such external checks, "S" becomes a placeholder with no rule-governed application. As he writes:

> Let us imagine a table, something like a dictionary, that exists only in our imagination… If the mental image of the timetable could not itself be tested for correctness, how could it confirm the correctness of the first memory?
>
> (Wittgenstein 2009, §265)[18]

This is not merely about the impossibility of establishing normativity in isolation but about the impossibility of establishing *meaningful normativity* in isolation. A scientist who declares, "I observed X," but provides no operational definition of "X," no description of the instrument, no calibration standard, and no context, utters a sentence that is cognitively empty.

The theory of theory-ladenness philosophy of science echoes the impossibility of private meaning. According to theory-ladenness, even the most "direct" observation is mediated by public frameworks. Seeing is not passive reception but an active, trained interpretation shaped by conceptual schemes and communal norms. Take the radiologist reading an X-ray. To the untrained eye, the image is a blur of grays. But the radiologist sees a tumor—not because their eyes are biologically superior, but because they have been socialized into a diagnostic community that shares criteria for identifying pathological structures. Without access to that shared framework, the image remains noise. When scientists "see" electrons in a cloud chamber, they do so not by pure perception but by trusting a lineage of stabilized practices: instrument design, statistical thresholds, theoretical models, and communal agreement about what counts as a "track." Meaning emerges not from purified individual vision but from material and social networks of verification.

Now consider the second sub-question. However, even if we grant that meaning is inherently public—that linguistic content, conceptual frameworks, and the interpretation of observations all depend on shared practices—it remains tempting to ask: Can an individual who *has already internalized this public background*, say, by respecting standardized procedures, achieve objective understanding entirely on their own? To put the question more concretely: Can a scientific claim be verified by a single person acting in isolation?

Intuitively, the answer seems to be "no." In contemporary scientific practice, it is now a matter of common sense that findings must undergo peer review. This requirement does not deny that a solitary individual might, in principle, arrive at conclusions that accurately reflect reality. Rather, it underscores a crucial epistemic and social point: until those findings are vetted and endorsed by multiple competent researchers, and a broader scientific community, the public will neither trust nor accept them, regardless of the relationship between the individual's claims and world.

The limitation here is not cognitive but testimonial: a single individual, no matter how meticulous, cannot by themselves generate the kind of warranted trust that underpins objective scientific knowledge in a communal enterprise. Indeed, contemporary scientific practice treats peer review as a necessary norm and this norm is often understood from a perspective on credibility which arises only when findings are subjected to scrutiny by multiple competent peers. Until a result is evaluated, replicated (at least in principle), and endorsed by others within the relevant epistemic community, it remains epistemically provisional.

That said, the notions of "trust," "public acceptance," and "credibility" invoked here are often left vague—and, importantly, they seem frequently mind-dependent, shaped by social expectations, institutional norms, or psychological dispositions. This raises some deeper questions: What exactly is lacking when a finding is endorsed by only a single individual? Is it a shortfall in reliability? A deficit in verifiability? Or something more fundamental about the conditions under which understanding—and knowledge—can arise at all?

The next chapter will address this question through the lens of Donald Davidson's argument on the problem of objectivity. For now, the key takeaway is this: objectivity is not a property of isolated minds, but of practices embedded in a public world. From Wittgenstein's rejection of private language to the widely accepted traditions in modern scientific practice, the message is consistent: there is no private path to objectivity.

Notes

1 Popper, Karl. *Conjectures and Refutations: The Growth of Scientific Knowledge*. New York: Harper, 1962.
2 Nissenbaum, Helen. *Privacy in Context: Technology, Policy, and the Integrity of Social Life*. Stanford, CA: Stanford University Press, 2009.
3 See Aristotle. *Politics*, 1253b, 1259b. Translated by Ernest Barker, revised by Richard Stalley. Oxford: Oxford University Press, 1995.
4 Carnegy-Arbuthnott, Hannah. "Privacy, publicity, and the right to be forgotten." *Journal of Political Philosophy*, 31 (September, 2023): 494–516.
5 Rumbold, Benedict and James Wilson. "Privacy rights and public information." *Journal of Political Philosophy*, 27 (March, 2019): 3–25.
6 Husserl, Edmud. *Cartesian Meditations: An Introduction to Phenomenology*. Translated by Dorion Cairns. Hague: Martinus Nijhoff, 1991.
7 Crossley, Nick. *Intersubjectivity: The Fabric of Social Becoming*. London: Sage Publications, 1996.
8 Crossley, Nick. *Intersubjectivity: The Fabric of Social Becoming*. London: Sage Publications, 1996.
9 Reich, Wendelin. "Three problems of intersubjectivity—and one solution." *Sociological Theory*, 28 (March, 2010): 40–63.
10 Munoe, Paul. "Intersubjectivity." In *The Blackwell Encyclopedia of Sociology*. Edited by George Ritzer and Chris Rojek. Oxford: Wiley-Blackwell, 2019.
11 Munoe, Paul. "Intersubjectivity." In *The Blackwell Encyclopedia of Sociology*. Edited by George Ritzer and Chris Rojek. Oxford: Wiley-Blackwell, 2019.
12 Popper, Karl. *The Logic of Scientific Discovery*. London and New York: Routledge, 2002.
13 Popper, Karl. *The Logic of Scientific Discovery*. London and New York: Routledge, 2002.
14 Megill, Allan. "Four senses of objectivity." In *Rethinking Objectivity*. Edited by Allan Megill, 1–29. Durham: Duke University Press, 1994.
15 See, Longino, Helen. *The Fate of Knowledge*. Princeton, NJ: Princeton University Press, 2002.
16 See Burge, Tyler. *Origins of Objectivity*, 46–49. Oxford: Clarendon Press, 2010 for more discussions on objectivity on subject matter.
17 Wittgenstein, Ludwig. *Philosophical Investigations* (4th ed.). Translated by Gertrude Anscombe, Peter Hacker, and Joachim Schulte. Oxford: Wiley-Blackwell, 2009.
18 Wittgenstein, Ludwig. *Philosophical Investigations* (4th ed.). Translated by Gertrude Anscombe, Peter Hacker, and Joachim Schulte. Oxford: Wiley-Blackwell, 2009.

3 The Problem of Publicity

In Chapter 2, publicity is treated as a necessary, or minimal, condition for scientific objectivity. Importantly, publicity here is not understood merely as a procedural norm or a social convention. It is a substantive epistemic precondition. Yet the path from private cognition to public knowledge is far from straightforward. For each individual, scientific cognition begins in subjectivity: in private experience, embodied perception, and frameworks shaped by culture, language, and history. This raises a fundamental tension: if all knowing starts from a perspectival, private standpoint, how can it ever yield knowledge that is genuinely public—that is, interpersonally valid and anchored in a shared world?

This tension gives rise to a two-part question, both strands of which have deep roots in traditional philosophy but have been largely neglected in contemporary discussions of scientific objectivity in the philosophy of science.

The possibility of solitary objectivity: Can a competent, honest individual or even a single research group working in complete solitude achieve objective knowledge? Modern scientific practices such as peer review, replication, and open data sharing reflect a deep-seated skepticism toward the epistemic self-sufficiency of any isolated agent. Does this skepticism make sense?

The genesis of publicity: If objective knowledge requires publicity, then how can inherently subjective starting points, private perceptions, personal observations, ever give rise to representations that are genuinely interpersonally accepted (and even about the same world)?

These questions are deeply intertwined and closely connected to the classical philosophical problems of objectivity and other minds. In the previous chapter, I briefly addressed the first question from the perspective of the public nature of meaning, drawing on Wittgenstein. In what follows, I further explore this issue, primarily within the context of scientific objectivity, by turning to Donald Davidson's account of objectivity, especially his triangulation thesis. Moreover, I argue that an aspect related to the triangulation approach to objectivity has been overlooked in existing discussions, yet it bears directly on the problem of publicity.

DOI: 10.4324/9781003783787-4

Gaps in Varieties of Knowledge

Regarding the possibility of solitary objectivity, that is, whether a solitary individual can independently attain objective knowledge, Donald Davidson offers a negative answer. His argument unfolds in three interlocking steps and culminates in a reconceptualization of how the notion of objectivity itself arises.

Step 1: Three Varieties of Knowledge

Davidson distinguishes three fundamental kinds of knowledge that human beings possess:

Knowledge of one's own mental states (e.g., "I believe it is raining");
Knowledge of the physical environment (e.g., "It is raining outside");
Knowledge of other minds (e.g., "She believes it is raining").

While these three forms are deeply interconnected, they differ in their modes of access. A subject—call her "I"—has what appears to be direct, non-inferential access to her own mental states. No empirical investigation is required to know what one believes or feels. Knowledge of the external world, by contrast, arises through sensory interaction with the environment: light, sound, pressure, and other stimuli give rise to perceptual beliefs about objects, events, and states of affair. Though mediated by the senses, this environmental knowledge is still, in a sense, "first-personally" available—it is my perception of rain, my belief formed in response to sensory stimuli.

Crucially, however, knowledge of other minds is not directly accessible. Unlike my own thoughts or my thoughts on my immediate surroundings, others' mental states cannot be observed from the inside. Even behavioral cues (e.g., facial expressions, utterances, gestures) are open to interpretation and require inference. In cases where such cues are absent, access to their mental life vanishes entirely (Davidson 2001, 205–221)[1].

Step 2: Davidson's Critique of Traditional Epistemology

Davidson argues that since Descartes, mainstream epistemology has treated subjective immediacy as the secure foundation of all knowledge. For Descartes, this foundation was the indubitable awareness of one's own existence and mental states (*cogito, ergo sum*). For Locke, Berkeley, and Hume, it was the stream of sensory impressions or sense-data, the raw materials from which all concepts and beliefs are supposedly constructed. Both traditions share a common assumption: objective knowledge must be built upward from subjective certainties. But Davidson identifies a critical flaw in this picture. He argues that philosophers have overwhelmingly focused on the problem: How can we move from subjective experience to knowledge of an objective world? Yet they have largely ignored a more fundamental problem: How do people even acquire the concept of objectivity in the first place? Davidson insists that before discussing how objective knowledge is achievable, one must first possess a concept of objectivity (Davidson 2004, 3–18)[2]. Only with the concept of objectivity can one address how to obtain objective knowledge. Say, to speak of a "mind-independent world," to distinguish between appearance and reality, or to claim that scientific theories aim at truth rather than mere coherence, all of these presuppose *a prior* grasp of what it means for something to be *objectively true*. Indeed, taxonomies of the notions of objectivity in earlier discussions (e.g., notions of objectivity in ontological, methodological, or intersubjective senses) also typically take the notions as given, *treating them as a target for classification rather than as cognitive achievements needing explanation*.

The deeper question, then, is this: If all an individual ever directly accesses are subjective states ("I think," "I see," "I feel," etc.), how does the idea of an objective truth that transcends those states ever arise?

Step 3: Standard of Objective Truth Acquired Through Interpersonal Communication

Davidson's answer to the above question hinges on a profound insight about the nature of propositional thought. All knowledge, whether about inner feelings or outer environments, is expressed in propositions, and propositions are inherently truth-evaluable. To judge that a proposition is true, one must possess a standard for distinguishing truth from falsity, or objective from non-objective. Only with the concept of objective truth can one assess whether judgments about mental states or the physical environment are true or false, objective or non-objective (Davidson 2004, 3–18)[3]. But where does this standard come from? How is the concept of objective truth acquired? Davidson argues that a subject *cannot* determine whether her representation of the physical environment or objects is objective without communicating with other subjects about their shared physical environment or objects. Because objective truth depends on intersubjective communication, there is no private sense of objective truth, but only a public one. As Davidson puts it:

> All propositional thought, positive or skeptical, of the inner or of the outer, requires possession of the concept of objective truth, and this concept is accessible only to those creatures that are in communication with others. Knowledge of other minds is thus basic to all thought. But such knowledge requires and assumes knowledge of *a shared world of objects in a common time and space*. Thus the acquisition of knowledge is not based on a progression from the subjective to the objective; it emerges holistically, and is interpersonal from the start.
>
> (Davidson 2004, xv, 18)[4]

This marks a decisive break from Cartesian and empiricist models. For them, knowledge begins in the private theater of the mind and only later reaches out to others and the world. For Davidson, there is no "private theater" capable of generating genuine propositional thought. Propositional thought, and thus objectivity, requires a public stage.

He reinforces this point by drawing on Wittgenstein's rejection of private language. Just as a word whose meaning is fixed solely by private sensations cannot sustain norms of correct usage, so too a belief held in total isolation cannot be assessed for truth. As Tyler Burge notes, having a belief does not entail knowing its content with sufficient clarity to judge its truth. To know what it would mean for "*F* is *G*" to be true, one must know what counts as *F* and what counts as *G* (Burge 2010, 265–266)[5], and that knowledge is acquired through participation in shared linguistic and perceptual practices.

Thus, Davidson argues that knowledge is not constructed, aggregated, or induced from independent subjective knowledge but begins with an intersubjective, public standard of truth acquired through communication. Specific subjects then use this public standard to judge whether their knowledge or thoughts (e.g., about their physical environment) are true.

"Triangulation" and the Public Nature of Objectivity

If objectivity cannot arise in solitary, how, according to Donald Davidson, do multiple subjects jointly acquire the concept of objective truth? Davidson's answer lies in his "triangulation thesis." He states:

> Rather, the objectivity which thought and language demand depends on the mutual and simultaneous responses of two or more creatures to common distal stimuli and to one another's responses. This three-way relation among two speakers and a common world I call "triangulation." In the end, the idea is as simple as that of *ostensive learning*, but with an insistence that triangulation is not a matter of one person grasping a meaning already there, but a performance that (when fully fleshed out) bestows a *content* on language.
>
> (Davidson 2001, xv)[6]

Davidson further elaborates:

> The basic situation is one that involves two or more creatures simultaneously in interaction with each other and with the world they share; it is what I call triangulation. It is the result of a threefold interaction, an interaction which is twofold from the point of view of each of the two agents: each is interacting simultaneously with the world and with the other agent. To put this in a slightly different way, each creature learns to correlate the

> reactions of other creatures with changes or objects in the world to which it also reacts. One sees this in its simplest form in a school offish, where each fish reacts almost instantaneously to the motions of the others.
>
> (Davidson 2001, 128)[7]

Davidson's triangulation involves three essential relata: two (or more) subjects and a single shared object or feature of the physical environment. As illustrated in Figure 3.1—where S_1 and S_2 represent two distinct subjects and O denotes their common object or environmental stimulus—objective knowledge arises only at the intersection of two *simultaneous forms of interaction* "*in a common time and space*." On one hand, each subject responds to the same external object O (represented by the lines between S_1 and O, and between S_2 and O; those terminating at O and O itself are rendered as dashed lines to signify that, due to increased distance, the images of the object or stimulus might become blurred.); on the other, they communicate with each other and interpret one another's responses to O (the downward arrows beneath S_1 and S_2 represent each subject's response to the shared stimulus, while the two horizontal arrows between S_1 and S_2 represent the communications between the two subjects). It is this dual responsiveness to both the world and to fellow interpreters that makes objective knowledge of the world possible. Crucially, neither subject–object interaction in isolation nor subject–subject communication alone suffices. Only their interplay enables subjects to distinguish how things appear to them from how things actually are, thereby establishing the very possibility of objective knowledge about the world.

When different subjects engage in "triangulation," each subject interacts with the physical environment or objects (subject–object interaction) while simultaneously interacting with other subjects regarding the outcomes of their respective subject–object interactions (interpersonal interaction). The concept of objectivity or the standard of objective truth is established through these simultaneous subject–object and interpersonal interactions.

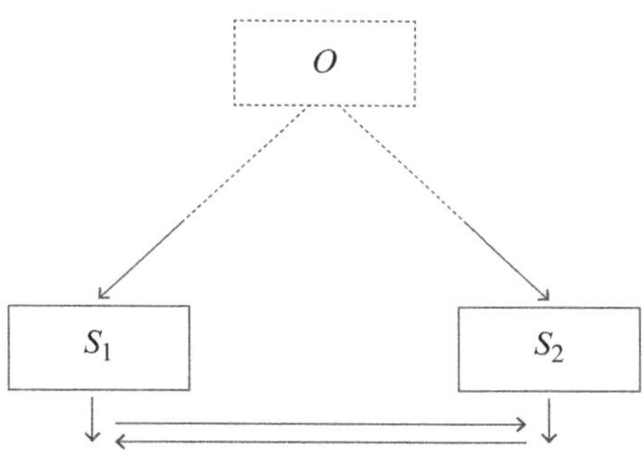

Figure 3.1 Donald Davidson's triangulation

While the fish might coordinate without concepts, human triangulation involves propositional attitudes—beliefs, assertions, and interpretations that are truth-evaluable. Crucially, it is through this interpersonal dynamic that subjects begin to recognize that their own representations might be mistaken, and that there exists a common reality against which such errors can be measured.

Davidson's triangulation argument can be unpacked from two complementary perspectives: (1) the publicity of meaning, and (2) the necessary conditions for objective representation. Together, they reveal why objectivity is not a private achievement but a public institution.

The Publicity of Meaning

One perspective of understanding Davidson triangulation thesis is from the lens of the philosophy of language and the publicity of meaning. From this perspective, the thesis functions as an argument against the possibility of a private language—akin to Wittgenstein's famous critique. Davidson's core claim is that thought is inherently linguistic, and language is inherently public, not merely in practice, but in principle. Crucially, he argues that objective meaning cannot be fixed by a solitary subject alone; it requires the interpersonal structure of triangulation.

For Davidson, language is not simply a collection of sounds or symbols. It is a structured system of expressions governed by syntactic rules and semantic interpretation:

> A language may be viewed as a complex abstract object, defined by giving a finite list of expressions (words), rules for constructing meaningful concatenations of expressions (sentences), and a semantic interpretation of the meaningful expressions based on the semantic features of individual words.
>
> (Davidson 2001, 107)[8]

This abstract architecture only becomes operationally meaningful—only acquires determinate content—through interpretive practice. When two subjects communicate about a shared object, they do far more than exchange utterances: they test hypotheses about each other's beliefs, calibrate reference, and gradually converge on stable patterns of usage. It is precisely within this dynamic three-place relation, "as simple as that of ostensive learning," in Davidson's words, between two interpreters and a shared feature of the environment that the meaningful content of language takes root.

From this perspective, meaning cannot be fixed by inner experience alone nor by one subject alone. To interpret another's utterance or even to stabilize the meaning of one's own thoughts, one must appeal to publicly accessible criteria grounded in shared behavior and environmental interaction. The significance of interpersonal communication is to put people in a position to make an objective distinction between correct and incorrect responses to external stimulus, and, as Davidson says, it bridges "the gap needed to make sense of the notion of error" (Davidson 2001, 4)[9]. In triangulation, all interlocutors establish *correlations* between their own responses to features of the environment and those of others.

That said, this view has not gone unchallenged. Some philosophers question whether triangulation is truly necessary for fixing meaning at all, for instance, arguing that meaning can be stabilized through other forms of social or causal interaction, without requiring the full-blown interpersonal structure Davidson describes[10].

Moreover, much of our everyday communicative practice, precisely because it is meaningful and effectively fosters interpersonal interaction, appears to function perfectly well without appealing to objective truth. Commands, questions, exclamations, fictional narratives, and expressive utterances are linguistically rich and fully meaningful, yet they are *not* truth-evaluable in the realist sense.

Similarly, many foundational forms of communication other than language, such as coordinating joint actions, expressing pain, or sharing affective states, often precede, and indeed help lay the groundwork for, the later emergence of objective thought. In early language acquisition or cross-cultural encounters, individuals routinely achieve mutual understanding through gestures, vocal intonations, facial expressions, and shared attention, long before engaging in truth-conditional judgments.

In short, meaningful discourse seems possible without appealing to objective truth. Storytelling, ritual speech, emotional expression, and collaborative play rely on empathy, practical coordination, or shared context, not on agreement about facts. This suggests that objectivity may not be a universal prerequisite for all forms of mutual understanding.

I partially agree with these possible critics. Fortunately, my own focus in this book lies on objectivity in scientific cognition, a domain that operates very differently from the non-truth-evaluable forms of communication discussed above. Also, I think Davidson's concern is narrower. He is not primarily addressing casual conversation or expressive speech acts; rather, he focuses on propositional attitudes—judgments that purport to represent states of affairs and are thereby subject to truth or falsity. As his distinction of three types of knowledge illustrates, such judgments encompass claims about one's own mental states, the external world, and other minds. Davidson insists that all three constitute forms of representational knowledge, and all presuppose a shared standard against which their accuracy can be assessed.

Consequently, while everyday communication may tolerate ambiguity, metaphor, or non-referentiality, the capacity to make and understand truth-evaluable claims in scientific practice demands a framework of objective truth. And that framework, Davidson argues, is only possible through triangulation. As he writes:

> we want to understand the actual utterances of others, and we want our utterances to be understood. What has language to do with this interest? The answer is that it is only by employing such concepts as word and sentence that we can give a systematic description of the linguistic aspects of linguistic behavior and aptitudes…
>
> (Davidson 2001, 109)[11]

In sum: Even if communication in general need not presuppose objectivity, but the understanding of propositional content, and the very possibility of error, does.

Triangulation as a Necessary Condition for Objective Representation

Beyond the philosophy of language, Davidson's triangulation argument may also be understood as addressing the necessary conditions for objective representation of the physical environment. The three types of knowledge he distinguishes, knowledge of one's own mental states, knowledge of the external world, and knowledge of others' mental states, all aim to represent states of affairs. Specifically, knowledge of one's own mental states represents the internal state produced in the mind caused by physical stimuli; knowledge of the physical environment seeks to represent the state of the same stimuli or surrounding world; knowledge of others' mental states attempts to represent the states produced in other minds by the same physical environment or objects.

For Davidson, a solitary subject cannot determine whether her representation is objectively true. To assess whether a representation is accurate or whether it qualifies as a good, objective representation, one requires an external standard. This is precisely where triangulation is introduced: objectivity functions as the criterion for evaluating *representational success*. As Tyler Burge interprets, Davidson aims to show that

> For an individual to have representational states[,] thoughts or perceptions with definite representational content there must be some non-arbitrary fact or ground that fixes what those thoughts are about … having a concept of objectivity is supposed to be necessary for representation of, and as of, the physical environment.
>
> (Burge 2010, 269)[12]

Thus, both objective thoughts and the very concept of objectivity emerge through communicative interactions among subjects about their beliefs concerning the world. On one hand, when two subjects are situated in the same physical environment or attend to the same object, they use language to interpret and translate each other's utterances about their respective beliefs. Through this process, they establish correspondence relationship between their belief systems, which allows them to actually interpret each other. On the other hand, achieving objective representation of the physical world itself requires that subjects communicate their individual representations of the shared environment, via triangulation, in order to *converge* on stable, intersubjectively validated content.

That said, while Davidson provides a compelling framework for how the concept of objectivity arises through triangular interaction, this framework remains notably vague, even underdeveloped, leaving some puzzles (see McDowell 2003; Montminy 2003; Verheggen 2013)[13]. After examining his argument, one is still left without a clear account of the specific standards that constitute objectivity, except that communications among subjects on the same stimuli are necessary for having a concept of objectivity. If objective truth depends on communication among subjects regarding the same physical environment or object, does it thereby depend on the group, the prevailing theoretical framework, or the dominant research paradigm, or even power structures?

Moreover, Davidson's triangulation model assumes certain basic conditions, for instance, that at least two subjects *simultaneously* share the same physical environment or attend to the same object in a common space. While this may seem commonsensically unproblematic, it raises complications from the standpoint of modern physics. Special relativity, for example, challenges everyday notions of simultaneity, showing that "simultaneous" events are reference-frame-dependent or subject-dependent. Also, even in mundane settings, say, two people in the same small room, their immediate perceptual and causal environments may not be identical. No two individuals can occupy the exact same location at the same moment or interact with precisely the same physical object. I will argue further later that the assumption of a fully shared environment is far less straightforward than it initially appears. Nevertheless, one point remains clear: objectivity, and objective truth, has an irreducibly public or interpersonal character. Regardless of how we ultimately specify its standards, the very concept of objectivity, as well as the criteria for objective truth, depends on communication among subjects and the public language that emerges from such interaction. It is through this shared linguistic medium that subjects can interpret one another's utterances (i.e., their beliefs about the world) and thereby align their representational frameworks.

In short, the concept of objectivity is not something a solitary mind can generate in isolation. It is collectively constituted through the interactions of at least two subjects. Consequently, a solitary subject cannot possess objective knowledge, not because their representations are necessarily false, but because the very notion of objective truth presupposes a public, interpersonal space of validation.

Problem of Publicity in Two Dimensions

Two Constitutive Problems

In previous discussions, I treated publicity as a necessary or minimal condition for notions of scientific objectivity in whatever sense. Yet this commitment immediately confronts a deep-seated tension between the subjective, private origins of all empirical cognition and the inherently public character of scientific knowledge as a collective, communicable achievement. This tension gives rise to two interlocking questions: the possibility of solitary objectivity, and the genesis of publicity on how do multiple individuals jointly arrive at representations that are not only mutually intelligible but also anchored in a common world.

If, as it has been argued earlier, a solitary individual, even one equipped with public concepts and methods, cannot attain objectivity, then the burden shifts: How is publicity achieved at all? I now turn to this second, constitutive question: How can inherently subjective starting points ever give rise to representations that are genuinely public?

To frame this inquiry more precisely, recall the definition of publicity: for any instance of scientific cognition to qualify as objective, it must, in principle, be accessible to *multiple individuals* and reflect a shared understanding of *a same subject matter*, one that transcends the confines of any individual's private experience. This does not require universal agreement across *all* observers, but it does demand that knowledge be open to scrutiny, interpretation, and corroboration by multiple individuals.

For this condition to be met, two interdependent conditions must be jointly satisfied:

Interpersonal intelligibility condition: Participants must be able to interpret one another's representations with sufficient fidelity, not merely decoding linguistic tokens, but grasping the conceptual roles, inferential commitments, and evidential functions embedded within them. Agreement here is not about surface consensus but about mutual recognition of meaning.

Referential identity condition: Those engaged in discourse must be talking about *the same thing* in the world. Their claims must refer to a shared subject matter, not just similar labels or overlapping descriptions, but a same object of inquiry whose properties and behaviors constrain their assertions.

Only when both conditions hold can people speak of a genuine shared understanding of the same object of inquiry. The failure of either condition undermines publicity, and thus objectivity, even if a superficial consensus appears to exist. Agreement on words without shared reference, and shared reference without mutual understanding, neither constitutes genuine agreements.

The Interpersonal Intelligibility Condition

The interpersonal intelligibility condition points toward what one might call a public sphere of meaning: a domain constituted by shared concepts, norms of inference, and linguistic practices that enable communication, criticism, and evaluation. Such a sphere cannot emerge if participants fundamentally disagree about how to interpret their own or others' concepts and representations. From this perspective, publicity is not a precondition assumed at the outset of inquiry but the epistemic *product* of successful scientific practice.

Yet, achieving interpersonal intelligibility faces two interrelated challenges.

Problem 1: The Problem of Other Minds and Public Meaning

How can one individual understand what another understands, given that the other's mind is not directly accessible? If one cannot verify—even indirectly—that others mean what one takes them to mean, how can one establish a stable framework for scientific discourse? My question on the possibility of solitary objectivity directly responds to this challenge. Within the context of the first condition for publicity, my discussions on Wittgenstein's view on the publicity of meaning and Davidson's triangulation thesis aim to show that publicity is not merely conducive to objectivity but constitutive of it. As Davidson argues, the very possibility of having propositional thoughts about an objective world requires a triangular relationship among at least two interpreters and a shared external reality. Without such publicity, there is no way to distinguish belief from mere sensation or error from truth. Thus, objectivity presupposes publicity.

Problem 2: Divergent Starting-Point Contexts

Even if one could resolve the problem of interpretive access to other minds, another obstacle remains: individuals may begin from radically divergent cognitive or philosophical foundations. They may endorse different ontologies (e.g., entities or processes), methodological priorities (e.g., prediction vs. explanation), or epistemic values (e.g., simplicity vs. empirical coverage). In such cases, disagreement is not a failure of communication but a constitutive divergence—a clash not over data, but over the very criteria that make data meaningful.

Thomas Kuhn's account of scientific paradigms is an example of this challenge. Scientists working within incommensurable paradigms, such as Greek astronomy and Newtonian astronomy, may observe the same skies, yet see different worlds. Their instruments may register identical numbers, but those numbers play distinct theoretical roles. Concepts like "mass," "time," or "element" shift meaning across paradigms, not as semantic drift but as reconfiguration of the entire conceptual field. Some of their disagreements are not empirical disputes resolvable by experiment; they are framework-level incompatibilities. In such contexts, the dream of universal, context-independent objectivity appears unattainable. But this does not invalidate publicity as a necessary condition for scientific objectivity. On the contrary, it refines it.

Kuhn himself emphasized that normal science operates within communities bound by shared paradigms. Within such communities, scientists achieve robust agreement on problems, methods, standards, and solutions. They can interpret one another's work, replicate experiments, and adjudicate disputes precisely because they inhabit a common epistemic world (Kuhn 1962, 35–51)[14]. This intra-paradigmatic consensus satisfies the publicity condition: multiple individuals jointly affirm the same claims about the same subject matter.

Thus, as long as we do not demand objectivity or publicity as universal agreement across all possible perspectives, divergent starting points may not pose an insurmountable obstacle. Interpersonally shared paradigms, along with mutually accepted or respected standardized procedures, can effectively "flatten" epistemic divergence within a group. Therefore, publicity as an indispensable baseline remains respected. In this sense, publicity, understood not as global consensus but as localized, interpersonal accountability, remains an indispensable baseline for scientific objectivity.

Referential Identity Condition

I now turn to the second condition for scientific publicity: referential identity. For scientific discourse to be genuinely public, participants' representations must be directed toward the same subject matter in the world. This condition is also explicitly articulated in Davidson's formulation of the triangulation thesis as "thought and language demand depends on the mutual and simultaneous responses of two or more creatures to *common* distal stimuli and to one another's responses" (Davidson 2001, xv)[15], although he takes this condition for granted and does not consider the possibility that the two creatures might, in fact, be responding to different stimuli.

Indeed, agreement in understanding carries epistemic weight only if it concerns a common object of inquiry whose properties and behaviors constrain the assertions made about it. If two researchers believe they are discussing the same phenomenon but are in fact referring to distinct (though perhaps superficially similar) entities, their apparent consensus is illusory. Such "agreement" does not reflect convergence in knowledge but rather a *mutual misunderstanding masked by linguistic overlap*. In these cases, claims may appear compatible on the surface, yet they offer no genuine mutual confirmation. Worse still, this superficial harmony can generate *a false sense of corroboration*: each party may interpret the other's findings as support for their own theory, when in reality they are investigating different things altogether.

The danger here lies not in overt disagreement, which at least signals a need for clarification, but in undetected referential divergence: a situation where different individuals thought they were talking about the same thing while reference silently drifts apart.

Fortunately, such misalignments are often self-correcting in scientific practice. When individuals are genuinely talking about different things, discrepancies tend to emerge or amplify over time, not necessarily in words, but in behavior, predictions, interventions, or experimental outcomes. These growing tensions eventually disrupt the illusion of consensus and compel a re-examination of what is actually being studied.

However, a more insidious scenario arises when the subject matters under discussion are *highly similar* in their observable properties. In such cases, participants may sustain fluent, apparently coherent communication for extended periods, even while operating with subtly different referents. Minor anomalies in data or interpretation are readily dismissed as experimental error, measurement noise, or acceptable theoretical uncertainty. As long as these discrepancies remain within tolerable margins, the participants may continue to treat its frameworks as compatible, unaware that they are, in fact, tracking distinct phenomena that happen to behave alike under limited conditions.

Only when accumulated anomalies exceed the threshold of what can be absorbed as "noise" does the illusion of shared reference begin to unravel. At that point, what once seemed like a minor deviation may reveal itself as evidence of a deeper referential mismatch, not a difference in opinion, but a divergence in the very object of inquiry.

This possibility highlights a crucial epistemic vulnerability: *publicity can be mimicked by similarity in subject matter*. Surface-level linguistic consensus, terminological consistency, and even predictive convergence are not, by themselves, sufficient proof of shared reference. True publicity requires not merely linguistic or behavioral alignment, but stable, verifiable coordination around a common worldly target, a condition that must be continually tested, not assumed.

This possibility is especially acute in sciences dealing with complex, historically contingent, or non-reproducible systems. The following are two simple examples on this possible in scientific research.

Example 1

Imagine two student researchers in two different cities, say Hangzhou and Chengdu, both located near 30°N latitude and participating in the same education project. Their task is simple: measure how the period of a pendulum changes with its length, testing the classic formula

$$T = 2\pi\sqrt{\frac{L}{g}},$$

where T is the swing time (period), L is the length of the string, and g is gravitational acceleration.

Both researchers follow the exact same protocol:

- Use a metal bob of 100 grams,
- Measure string length from pivot to center of the bob,
- Release the pendulum from a 10° angle,
- Time 10 full swings with a digital stopwatch, then divide by 10,
- Repeat for lengths from 20 cm to 100 cm in 20-cm steps.

After collecting data, both researchers plot T versus $\sqrt{L}$. The graphs are just as theory predicts. They calculate values of g around 9.78–9.81 m/s^2, close to the expected value. They publish their reports online, and teachers praise the experiments as a "successful confirmation of Newtonian physics." But here's what no one notices: the two observers aren't actually measuring the same physical system.

The Hangzhou researcher conducts her experiment on the third floor of a modern steel-frame school building near a Lake. Her pendulum hangs from a "rigid" lab stand bolted to a heavy table. The air is still, and she uses a thin, non-stretched nylon thread.

The Chengdu researcher works in an older brick schoolhouse with wooden floors. Every time a car rumbles past, her lab stand trembles very faintly—so imperceptibly that she hardly notices it. Unaware, she uses a cotton string that stretches very little under the weight of the bob, especially as the pendulum swings and slightly heats up from friction. Also, her lab has an air conditioner running to combat humidity, creating faint air currents.

These differences seem minor, certainly not enough to "break" the experiment. And indeed, both datasets look correct. But the underlying causes of the measured periods are not identical: in Hangzhou, the motion is *nearly* ideal: dominated mainly by gravity and string length; in Chengdu, the period is subtly influenced by string elasticity, air drag from the air conditioner, and tiny vibrations in the support structure, effects that mimic or mask small deviations from theory.

Now suppose a third group tries to use both datasets to test a refined model that includes air resistance. They're confused: the Chengdu data shows bigger disagreements, while Hangzhou's doesn't. Only then do they realize that although both researchers studied "a simple pendulum," the physical referents were different: one was close to the textbook ideal; the other was a real-world hybrid system where multiple forces quietly shaped the outcome.

Crucially, neither researcher made a mistake. Both followed mechanical instructions *faithfully* or are procedurally objective. Yet because complex real-world conditions can't be fully controlled, even in a "simple" experiment, their shared label ("pendulum period vs. length") pointed to two slightly different phenomena.

This illustrates a quiet truth in experimental science: reproducing a method does not guarantee reproducing an exactly same phenomenon. When "hidden" or "implicit" variables, material properties, environmental noise, or structural details, differ in ways that people don't measure or even notice, two accurate experiments may appear to confirm each other while actually describing distinct physical situations.

In classrooms and cutting-edge labs alike, this means that agreement in results isn't always evidence of understanding the same thing—only that the differences haven't yet mattered enough to show up. True reproducibility requires not just "copying" steps but awareness of broad and variable context, humility about assumptions, and openness to the idea that even the simplest systems carry invisible complexity.

Example 2

Similarly, in ecology which often deals with complex systems, a study on the foraging behavior of a bird species in a particular forest may be influenced by countless interacting variables, predator presence, seasonal food availability, microclimatic shifts, human disturbance. None of which can be *fully* controlled or replicated. Two researchers, both collecting *first-hand* observational data, might publish seemingly consistent findings using the same species name and methodology, yet be observing responses to subtly different ecological systems, thereby anchoring their claims to non-identical referents.

Consider a straightforward field project carried out by two researchers from different regions as part of a national biodiversity monitoring program. Both aim to answer the same question: "How does the distance from forest trails affect the ground-foraging activity of the red-billed leiothrix (Leiothrix lutea), a common songbird in subtropical forests?"

They follow an identical protocol:

- Select five forest sites within protected areas,
- Walk fixed 200-meter transects perpendicular to main trails at dawn,
- Record every sighting of a red-billed leiothrix foraging on the ground within 50 meters of the path,
- Repeat weekly for six weeks during early spring.

Both researchers find the same spatial pattern: ground-foraging birds are rarely seen within 10 meters of trails but become common beyond 30 meters. They conclude that "red-billed leiothrix avoids trail edges, likely due to human disturbance." Their reports are published in the same journal, and their results are often cited together as evidence of this species' sensitivity to human presence.

But unbeknownst to either researcher, the forests they studied, though both classified as "subtropical evergreen broadleaf," differ in ecologically significant ways that reshape what "avoidance" actually means.

In one site, the forest is surrounded by quiet rural settlements. Trails see only occasional foot traffic, mostly from local residents walking slowly. The real reason birds avoid the trail edge isn't people. It's free-roaming domestic cats that congregate near trailheads, hunting small birds. The leiothrix isn't reacting to noise or movement; it's evading a concentrated predation risk linked to human habitation.

In another site, the forest borders a popular nature reserve with frequent visitors (e.g., school groups, photographers, and weekend hikers). There are few cats, but constant noise, sudden movements, and airborne dust from heavy footfall. Here, birds avoid the edges because chronic sensory disturbance makes foraging inefficient and stressful, not because of direct predation.

Both researchers observed the same distribution pattern and used the exact same method. Their data are accurate and reproducible within their own contexts. Yet the underlying drivers differ: one reflects a response to predation pressure, the other to anthropogenic sensory stress. Because neither research measured variables like predator density, ambient noise levels, or behavioral indicators such as vigilance time, they assumed they were documenting the same phenomenon.

Only later, when a conservation agency applies their joint recommendation—"maintain 30-meter no-access buffers around trails to protect ground-foraging birds"—do discrepancies emerge. In one region, the buffer reduces cat access and successfully increases bird activity near trails. In another region, birds still avoid the area because visitors gather just outside the buffer zone, continuing to generate noise and movement. The identical protocol strategy fails there, not because the original data were wrong but because the ecological meaning of "trail avoidance" was never identical across sites.

This illustrates a fundamental challenge in studying complex systems like ecology systems: even careful, well-intentioned replication can conceal hidden heterogeneity or contextual differences. When living systems are embedded in unique local webs of species interactions and human influence, shared methods and taxonomic labels do not guarantee that researchers are studying the same phenomenon. True scientific understanding requires not just repeating procedures but continually asking: What is this pattern really responding to and could it mean something different elsewhere?

In other similar cases, the climate event that cannot be replayed, and the ecological interaction shaped by unrepeatable contingencies, the economic shock embedded in a unique historical conjuncture. The irreducible uniqueness of context means that verification cannot depend on exact replication. Instead, scientists must rely on alternative strategies: robustness across modeling frameworks, analogical extrapolation from related systems, statistical generalization over heterogeneous datasets, or convergent evidence drawn from multiple independent methods. These approaches are not second-best compromises; they are essential epistemic tools for navigating domains where controlled repetition is impossible.

Yet crucially, these very strategies *presuppose* referential identity. They operate under the assumption that, despite variation in context, measurement technique, or theoretical framing, researchers are still tracking the same underlying phenomenon, whether labeled "climate sensitivity," "trophic cascades," or "price elasticity." Robustness only counts as confirmation if the models being compared are *genuinely about the same thing*; analogies only illuminate if the source and target share relevant causal structures; and convergence across methods only strengthens belief if those methods are probing a common reality.

If this assumption fails—if researchers are unknowingly studying different natural kinds under the guise of a shared term—then even the most sophisticated validation techniques may not correct error but amplify and entrench it. Consensus built on misaligned reference becomes a form of collective illusion: a house of cards held together by linguistic convention rather than empirical anchoring. In such cases, methodological rigor can ironically deepen epistemic vulnerability, lending false confidence to conclusions that rest on a fractured foundation.

One might argue that, in the long run, differences in referents will eventually surface. But if this emergence takes too long, the original systems may have evolved significantly in the interim. This raises a further complication: when discrepancies finally appear, it may be difficult to determine whether they stem from (1) the changes in the original complex systems being studied, or (2) divergence between the original systems and their later reproductions or models. In such cases, the very criterion for detecting referential drift becomes ambiguous.

Thus, these examples underscore a broader lesson: in domains where exact replication is impossible, referential identity cannot be guaranteed by protocol alone. It must be actively maintained through ongoing calibration, cross-validation across contexts, and critical scrutiny of whether apparent consensus truly reflects engagement with the same phenomenon. Only then may the interpersonal intelligibility condition serve as a scaffold for genuine publicity.

The Irreproducibility of First-person Observations

Previous discussions on referential identity center chiefly on cases where distinct individuals or research groups study what seems to be the same phenomenon, albeit with subtle variations, yet still obtain highly convergent empirical results. The convergence of outcomes, combined with shared terminology and methodological protocols, leads different researchers to believe they are studying the same phenomenon. Yet, as being pointed out, subtle but consequential differences in their actual systems or boundary conditions mean that their investigations are anchored to non-identical referents. This mismatch does not constitute the kind of triangulation described in Donald Davidson's theory. In Davidson's framework, triangulation arises when two agents and a shared external stimulus jointly enable the emergence of objective content through interpretation and correction. But in these ecological or physical-science cases, there is no single shared stimulus; instead, there are similar but distinct physical systems that happen to produce analogous data under limited conditions. The publicity of representation here rests not on access to a common object, but on *a belief in the homogeneity of stimuli*, especially natural kinds.

Moreover, there is a second dimension of referential identity that introduces a deeper epistemic constraint rooted in the very fabric of physical reality. In this dimension, the issue is not that observers are observing different objects under different conditions, but rather that each act of observation constitutes a unique physical event: an irreducible interaction between the observer, conceived not as a disembodied mind but as a concrete physical system, and the target object, itself another localized physical system, occurring at a unique point in spacetime.

From the perspective of relativity theory, such an interaction is the intersection of two worldlines: one tracing the history of the observer (or their measuring apparatus), the other that of the observed entity. Crucially, each observation made by an observer is irreducible to any other observer's observation, because every observational act is a singular, non-duplicable event in spacetime. No two observers can occupy the exact same spatiotemporal location; thus, no two observations are strictly identical. Even when multiple scientists aim their instruments at "the same" particle, star, or pendulum bob, they each engage with it through distinct causal channels at different times, from different angles, via different mediating fields or detectors. Consequently, the events they record are, strictly speaking, distinct singular events.

Consider again the simple pendulum experiment. Two researchers, working independently in separate laboratories, follow an identical protocol: same pendulum bob mass, same string length, same release angle, and same timing method. They both measure a period of approximately same seconds and conclude that gravitational acceleration is about 9.8 m/s^2. Superficially, this looks like successful replication—and in many practical contexts, it is treated as such.

But physically, no two observations are ever identical events. Each measurement involves:

- A distinct pendulum: even if manufactured to the same specification, its string has microscopic variations in elasticity, its bob has unique surface oxidation, and its pivot point experiences slightly different friction.

- A distinct environment: one lab may be 0.003°C warmer, causing air density to differ; another may sit above a subway line whose vibrations subtly perturb the support structure or a barely perceptible air current.
- A distinct observer-system interaction: the way each student holds the stopwatch, their reaction time, even the gravitational field at their exact latitude and elevation—all contribute to the total physical context of the measurement event.

These differences are not merely "noise" to be averaged out; they are constitutive of the observation as *a spacetime event*. No two such events can be perfectly identical or "replicated," because perfect replication would require duplicating not just the nominal variables (length, mass, angle) but the entire local state of the environment or even the universe at the moment of measurement, including quantum fluctuations, thermal motions, and gravitation field distribution, etc.

This problem becomes even more acute in quantum mechanics. If the observed entities are electrons or photons, the particles deemed "indistinguishable" by quantum statistics, *the particles may be identical, but the measurement events remain distinct*. Each detection occurs at a unique spacetime point, mediated by a specific detector with its own calibration history, thermal state, and coupling to the electromagnetic vacuum. Thus, even when observing "the same particle type," the physical events of observation are never located in the exact same location or simultaneous; they are separate coincidences in spacetime.

Crucially, this challenge exists independently of the problem of other minds. Even if we grant full transparency of mental states, the physical basis of observation itself resists exact duplication. Every measurement is a localized, historically contingent interaction between two material systems under a unique set of boundary conditions.

Of course, in practice, scientists routinely ignore these "micro-differences." They assume that small fluctuations in temperature, humidity, or vibration are negligible—that the background is "flat" enough to treat repeated measurements as instances of the same phenomenon. But *this is not a discovery of facts; it is a methodological assumption of homogeneity*. It is a pragmatic decision to treat diverse spacetime events as equivalent for the sake of building stable or generalizable models.

Yet this assumption carries epistemic risk. When systems are highly sensitive to initial or boundary conditions, as in chaotic dynamics or some complex systems, those "negligible" differences can cascade into divergent outcomes. What appears to be replication may, in fact, be convergence under coarse-grained description, masking underlying heterogeneity.

Thus, the publicity of scientific representation cannot rest on the wishful belief that "we are all looking at the same thing." Instead, it must acknowledge that all observations are singular events, and that some of the shared understanding in scientific practices actually emerges not from identity of referents, but from robustness across variation—from the ability of a concept or model to hold up despite the inevitable uniqueness of every act of observation.

Now, let us return to Davidson's triangulation thesis again. As argued above, a crucial but often underemphasized implication of the physical nature of observation is that each act of perception or observation is an irreproducible spacetime event. When Subject 1 (S_1) observes an object O, and Subject 2 (S_2) observes what they both presume to be the same object O, they are not merely having different mental states or subjective perceptions. Rather, they are each participating in unique physical interactions, two distinct coincidences in spacetime, between their own embodied sensory apparatus (as a physical system) and the local manifestation of *O* under their respective environmental conditions.

As illustrated in Figure 3.2, S_1's observation occurs at one point in spacetime, shaped by S_1's position, motion, ambient temperature, lighting, gravitational field, and even quantum-level fluctuations in their measuring device. S_2's observation occurs elsewhere, under a different—but perhaps only slightly different—set of physical conditions. Even if both subjects use the same mechanical processes, the worldlines of observer and observed intersect only once per observation, and never in exactly the same way twice.

The key point is not that S_1 and S_2 have "different minds" or private inner experiences, an issue central to the traditional problem of other minds, but that their observations themselves, their respective immediate experiences, are physically non-identical events or "responses at the same time and location" as Davidson takes for granted. Each is a singular, non-repeatable encounter between a localized observer and a locally instantiated aspect of the world.

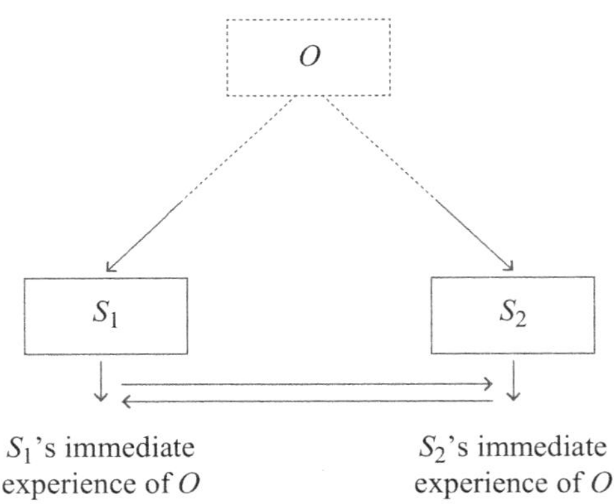

Figure 3.2 Triangulation with identity problem of immediate experience

This irreproducibility has a profound consequence: one observation cannot actually serve as direct corroboration of the other in the sense of referential identity. Corroboration in the sense of referential identity implies that two independent accesses converge on the same fact about the same entity. But if each observation is tied to a unique configuration of matter, energy, and context, then what appears to be convergence may instead be parallel responses to similar—but not identical—stimuli. There is no shared "stimulus" in the strict Davidsonian sense unless we already assume that the object O is stable and invariant across contexts—an assumption that itself requires justification.

If this line of argument holds, then the problem of publicity—how multi-subjective judgment emerges from private acts of perception—becomes significantly more complex. It is no longer enough to say that two agents respond to "the same stimuli" and thereby form public meaning through triangulation-like mutual interpretation. Instead, one must ask: How can different individuals achieve a shared understanding of a single reality when each is anchored to an irreproducible, context-bound observational event?

Consider a concrete example—the pendulum case once again. Two researchers, Alice and Bob, independently measure the period of a "50-cm pendulum" in the same laboratory at *nearly* the same time. Both obtain $T \approx 1.42\,s$ and agree: "Gravitational acceleration is about $9.8\,m/s^2$." From a practical standpoint, this appears to be a successful instance of triangulation: they refer to the same physical setup at (almost) the same time and place, and converge on a common conclusion.

Yet, strictly speaking, their observations are *not* of the same physical process but of two similar processes governed by the "same" laws yet embedded in subtly different boundary conditions. For instance, air density in the lab is not perfectly uniform or stable; the timer's or the pendulum's signal reaches Alice a fraction of a microsecond later than it reaches Bob; their reaction times differ slightly. Moreover, each observation reflects the interaction between the observer and their own slightly distinct local environment.

Despite these differences, Alice and Bob may still treat their experiences as instances of a shared stimulus—but only by appealing to an idealized theoretical model that abstracts away from such local contingencies. Through dialogue, they calibrate their interpretations, not merely by assuming access to a common external object but by actively aligning their conceptual frameworks around a stable theoretical construct. At the most fundamental level, then, Davidsonian triangulation does not strictly hold, especially not at the level of raw sensory input.

The agreement between Alice and Bob, accordingly, does not arise from the identity of immediate perceptual experiences, but from the stability of interpreted patterns across irreproducible events. The "shared object" is not given in perception; it is constructed through ongoing communicative and inferential practices that systematically treat local variations as noise relative to a hypothesized invariant structure.

In this view, publicity is not a given but a hard-won achievement based on a fragile consensus forged across a landscape of unique, unrepeatable encounters with a world that never presents itself identically twice. When we take into account both dimensions of the publicity problem—the epistemic gap posed by other minds and the referential gap arising from the non-identity of observational events—it becomes clear that achieving genuine multi-subjective agreement is far from straightforward. These gaps appear deeply entrenched: no two observers occupy the same spacetime point, and no two experiences are fully transparent to one another. From this perspective, Davidsonian triangulation, while insightful, seems insufficient on its own to bridge all these divides.

This is not to say that triangulation fails entirely. On the contrary, its core insight that meaning is inherently public, emerging only through interaction among agents directed at a shared world remains profoundly valuable. Even amid irreducible differences in immediate experience, the very public nature of meaning presupposes or builds in some form of common ground. Yet the challenge remains: If objective knowledge requires publicity, how can it arise from starting points that are inherently subjective, private, and singular events of perception?

This tension brings us to the central question of what might be called the genesis of publicity: How do representations that originate in individual, context-bound encounters come to be stabilized, shared, and recognized as being about the same world? In response, in the next chapter, I propose a transcendental approach to the constitution of publicity, one that seeks not to eliminate the gaps between referents, but to construct the necessary conditions under which such gaps can be traversed, however maybe imperfectly, in the pursuit of public understanding.

Notes

1 Davidson, Donald. "Three varieties of knowledge." In *Subjective, Intersubjective, Objective*. Oxford: Clarendon Press, 2001.
2 Davidson, Donald. "The problem of objectivity." In *Problem of Rationality*. Oxford: Clarendon Press, 2004.
3 Davidson, Donald. "The problem of objectivity." In *Problem of Rationality*. Oxford: Clarendon Press, 2004.
4 Davidson, Donald. "The problem of objectivity." In *Problem of Rationality*. Oxford: Clarendon Press, 2004.
5 Burge, Tyler. *Origins of Objectivity*. Oxford: Clarendon Press, 2010.
6 Davidson, Donald. "Three varieties of knowledge." In *Subjective, Intersubjective, Objective*. Oxford: Clarendon Press, 2001.
7 Davidson, Donald. "Three varieties of knowledge." In *Subjective, Intersubjective, Objective*. Oxford: Clarendon Press, 2001.
8 Davidson, Donald. "Three varieties of knowledge." In *Subjective, Intersubjective, Objective*. Oxford: Clarendon Press, 2001.
9 Davidson, Donald. "Externalisms." In *Interpreting Davidson*. Edited by Petr Kotatko, Peter Pagin, and Gabriel Segal, 1–16. Stanford, CA: CSLI Publications, 2001.

10 See Talmage, Catherine. "Meaning and triangulation." *Linguistics and Philosophy*, 20 (April, 1997): 139–145; Burger, Tyler. "Social anti-individualism, objective reference." *Philosophy and Phenomenological Research*, LXVII (November, 2003): 682–690.

11 Davidson, Donald. "The second of person." In *Subjective, Intersubjective, Objective*, 107–121. Oxford: Clarendon Press, 2001.

12 Burge, Tyler. *Origins of Objectivity*. Oxford: Clarendon Press, 2010.

13 Mcdowell, John. "Subjective, intersubjective, objective." *Philosophy and Phenomenological Research*, LXVII (November, 2003): 675–681; Montminy, Martin. "Triangulation, objectivity and the ambiguity problem." *Crítica: Revista Hispanoamericana de Filosofía*, 35 (December, 2003): 25–48; Verheggen, Claudine. "Trigangulation." In *A Companion to Donald Davidson*. Edited by Ernie Lepore and Krik Ludwig, 456–472. Oxford: Wiley Blackwell, 2013.

14 Kuhn, Thomas. *The Structure of Scientific Revolutions*. Chicago: University of Chicago Press, 1962.

15 Davidson, Donald. "Three varieties of knowledge." In *Subjective, Intersubjective, Objective*. Oxford: Clarendon Press, 2001.

4 Constitution of Publicity

I have argued in the last chapter that the problem of publicity remains unresolved, and specifically that Davidson's triangulation thesis cannot fully or genuinely secure publicity in science due to the problem of referential identity. In what follows, I propose an alternative approach to the genesis of publicity, developed from the standpoint of transcendental philosophy. To articulate this transcendental account of the constitution of publicity, just as Davidson distinguishes three types of knowledge, I begin by laying out a schema that differentiates types of scientific judgments. However, whereas Davidson's distinctions center on who has access to knowledge and who holds epistemic authority, my classification focuses instead on two features: the linguistic form of the judgments and the subject matter to which they refer. These distinctions may appear commonsensical at first glance, but they prove helpful for clarifying how publicity is constituted within scientific practice.

Types of Judgments in Scientific Discourse

Scientific judgments can be differentiated according to the nature of their content and the mode through which their contents are acquired. This yields a multi-dimensional classification that reflects the layered structure of empirical knowledge itself.

Judgments About Particular Events vs. Judgments About Universal Regularities

Scientific knowledge is typically expressed through propositions that assert something about the world (e.g., phenomena, regularities, or physical systems). Depending on their referential scope and temporal-spatial anchoring, these propositions fall into the following two broad categories.

DOI: 10.4324/9781003783787-5

Judgments About Particular Events

These judgments concern specific, spatiotemporally located occurrences, that is, events that have happened, are happening, or are predicted to happen under *determinate conditions*. Such judgments are always anchored in a concrete observational or experimental context. For example: "At 08:30 UTC on November 11, 2025, the period of a 1-meter pendulum in this vacuum chamber V at this Laboratory L was measured as 2.006 seconds by me."

A judgment of this kind relies on some subject's interaction with the world under a specific set of conditions at a particular point in spacetime. The event is not merely "out there"; it is known through a localized act of measurement, which is an interaction between observer, instrument, and physical system.

Judgments About Generalized Universal Regularities

These judgments do not refer to any single occurrence but express patterns, laws, or functional relationships that hold across multiple instances or under a range of conditions. They abstract away from the particularity of individual observations. For example: "The period of a pendulum is independent of its mass but proportional to the square root of its length." Such judgments may originate from repeated observations of similar particular events, from theoretical derivation, or from efforts to unify disparate phenomena. Their generality manifests in at least three interrelated dimensions:

Generalization across subjects: It is not just I who observes this relationship, but any competent observer, under relevant conditions, would obtain the same result.

Generalization across conditions: The regularity holds not only in one lab at one time, but whenever the specified physical conditions are met.

Generalization across objects: The judgment applies to a class of systems (e.g., all pendulums of a given length), but not just a single instance.

It is crucial to note that this distinction operates at the level of epistemic content, not ontology. It is not about whether "pendulumness" or "periodicity" exists as a Platonic universal, but rather how scientific judgments refer to events, either as singular, situated occurrences or as repeatable, law-like patterns.

Judgments of Direct or Immediate Experience vs. Judgments Derived From Immediate Experience

This distinction concerns how the content of a judgment is acquired, whether it stems from immediate engagement with the world or derived from immediate experience of the world.

Judgments of Direct or Immediate Experience

These "direct judgments" are grounded in a subject's first-hand, non-inferential encounter with a physical system, whether through unaided perception or instrument-mediated observation. They report what *this* subject, at *this* time and place, under *these* physical conditions, encountered. Their epistemic warrant does not derive from generalizations or theoretical inference, but from a direct causal interaction between the subject and the object of observation. For example: "At 6:12 a.m. on April 7, 2025, standing beside this chamber V, I saw this timer displays 2.006 s as the 1-meter pendulum completed its tenth oscillation." Such judgments are irreproducibly tied to a specific observer's immediate experience. Their authority arises from the concrete, spatiotemporally localized interaction among the observer, the measuring instrument, the surrounding physical environment, and the pendulum itself.

Judgments Derived From Immediate Experience

These judgments do not consist of immediate observational reports; rather, they are formed through inferential reasoning, generalization, or the testimony from other observers. They fall into several subtypes:

Inductive generalization: "After many observations, I conclude *all* pendulums of this length will always have this period."

Predictive inference: "Given the known length of a planned pendulum, the local gravitational acceleration $g = 9.81$ m/s^2, and the small-angle approximation, I predict that once built, it will have a period of $T = 2\pi\sqrt{L/g} \approx 2.0$ s."

Theoretical derivation: "Using Newtonian mechanics and a direct measurement of local g, I calculate that a pendulum of length 0.25 m will have a period of exactly 1.0 second, even though I have not yet constructed it."

Self-based derived judgments: A subject draws conclusions based on their own earlier direct observations. For example, after personally measuring the periods of several pendulums of varying lengths, a researcher infers the functional relationship $T \propto \sqrt{L}$ and applies it to design a new timing device.

Other-based indirect judgments: A subject incorporates the direct judgments of others, via testimony, published data, or communication, into their own reasoning. For instance, scientist A measures the period of a pendulum in a vacuum chamber and publishes the result. Scientist B, who has never performed the experiment, reads A's paper and uses the reported value to calibrate a gravimeter. For B, A's measurement functions as an indirect experience, but not personally enacted by B.

Even seemingly joint observations reflect this structure. For example, when two researchers say, "We observed the pendulum's period together," each still relies, in part, on the other's perceptual report as corroborating evidence. Their shared conclusion integrates both direct and indirect epistemic inputs.

Thus, knowledge derived from immediate experience, whether one's own or another's, extends the reach of inquiry beyond the limits of any single observer's time, location, or perceptual access. Of course, this extension comes at a cost: it introduces layers of interpretation, reliance on trust, and vulnerability to error that are absent in purely unmediated observation.

First-Person and Third-Person Judgments

From a linguistic and narrative perspective, judgments can be classified as first-, second-, or third-person based on the pronouns they employ and the perspective from which they are articulated. First-person judgments take an "I-form" or "We-form," such as "I think…," "I believe that…," "We observe…," or "We find…" Second-person judgments adopt a "You-form," for example, "You claimed that…" Third-person judgments refer to individuals by proper name (e.g., Galileo), definite description (e.g., the observer), or impersonal pronouns (he, she, it, they), and often omit explicit subject altogether, as in "It was observed that…"

While second-person narration, in which the reader is directly addressed as "you," is common in instructional texts, interactive fiction, and informal communication such as emails or lectures, it plays only a minor role in formal scientific writing. Scientific discourse primarily operates within first-person and third-person frameworks, each of which carries distinct epistemic implications. In what follows, I focus on these two modes, examining not only their grammatical forms but also their underlying epistemological differences.

First-Person "I-form" Judgments

A first-person "I-form" judgment explicitly situates the speaker within the act of knowing. It takes forms such as "I observe…," "I conclude…," or "I measured…," thereby anchoring the claim to a concrete subject's experience, reasoning, or action. Although this subject's personal perspective may align with the views of others or be shaped by a collective framework, such as a theory-laden interpretive scheme, a shared experimental procedure, or a prevailing paradigm, it remains fundamentally rooted in the individual's epistemic standpoint.

Such judgments are rooted in a specific, embodied speaker who exists in a definite physical and social context. The speaker is *a real individual*, situated in *specific* time and space, using particular instruments, operating under *specific* background conditions (e.g., temperature, lighting, theoretical assumptions), and interpreting data through her own perceptual and cognitive apparatus. Consider, for example: "I measured the period of my pendulum (length = 1.0 m, steel bob, released from 7°) as 2.01 s using a photogate timer in my lab at 22°C on September 15, 2025." This statement is not only about the pendulum; it is also *indexical*, referring directly to the speaker's unique encounter with the world and *inseparable from the specific conditions under which that encounter occurred*. It carries implicit limitations: what I saw, with my setup, at that moment, shaped by the sensitivity of my instruments, the ambient environment, and my own perceptual and interpretive framework. The "I" here is not egotistical; rather, it functions as an epistemic marker, signaling the situated origin of the data and acknowledging that empirical knowledge arises not in abstraction, but through concrete, embodied acts of observation.

However, "I-form" judgments carry characteristic epistemic risks. On one hand, a researcher may, even in good faith, mistake accidental features of her local context, such as a lab-specific calibration error or ambient mechanical vibration, for essential aspects of the phenomenon under study. On the other hand, certain background conditions, like the local gravitational field or atmospheric pressure, may be so "ordinary" or "unremarkable" to the observer that their contingent and variable nature goes unnoticed. What appears to be "*the way things are*" may in fact be no more than "*the way things are for me, here, now*," a perspective shaped by the particularities of one's immediate physical and experimental situation. This subtle conflation of the local with the universal constitutes a persistent challenge in empirical inquiry, precisely because it can remain invisible to the very subject or even a group of subjects whose testimony grounds the observation.

Of course, "I-form" judgments are not inherently unscientific. On the contrary, they constitute the raw material of empirical inquiry. Empirical knowledge invariably begins as an "I-judgment," a claim rooted in a singular, embodied encounter with the world, articulated by a situated observer who sees, measures, or records something at a specific time and place. Far from being a flaw, this first-person grounding is epistemically indispensable: it is through such concrete, localized acts that data enter the scientific record.

First-Person "We-Form" Judgments

Linguistically, "we-form" judgments, which employ the pronoun "we," also belong to the first-person category. Yet they are epistemologically distinct from "I-form" judgments in scope, authority, and the kind of epistemic responsibility they express.

In contrast to an "I-judgment" which is a personal, often private evaluation grounded in an individual's perception, interpretive framework, and cognitive biases, and which carries a clear sense of the individual "I"'s authorship (the speaker explicitly stands behind the claim as her own), a "we-judgment" expresses a collective epistemic stance or authorship. This "we" may refer to a concrete research team (e.g., co-authors of a paper), a disciplinary community (e.g., cosmologists), or an idealized community of rational inquirers committed to common standards of evidence, method, and logic. For instance, some cosmologists might jointly assert: "We conclude that dark matter constitutes approximately 27% of the universe's mass-energy content." This judgment does not merely report a personal belief; it asserts a claim that purports to be validated through communal practices, such as peer review, reproducible experiments, theoretical coherence, and consensus-building over time.

Notably, even single-authored scientific papers routinely adopt the "we-form." For example, Albert Einstein opens his 1948 paper "A Generalized Theory of Gravitation" with:

> In the following *we* shall give a new presentation of the generalized theory of gravitation, which constitutes a certain progress in clarity as compared to the previous presentations.
>
> (Einstein 1948)[1]

Here, "we" signals that the author does not wish to just speak as an isolated individual, a unique person always situated within particular contexts. Instead, the speaker implies that the judgment could, in principle, be replicated or endorsed by multiple researchers or even any competent researcher working under similar conditions, or *an aspiration* that the claim ought to be accepted by the scientific community once it has undergone proper scrutiny, even if full consensus has not yet been achieved.

Thus, while both "I" and "we" are grammatically first-personal, they represent or indicate different epistemological or psychological roles. "I" foregrounds the situated, embodied speaker. "We" gestures toward the ideal of communal knowledge[2], even when that ideal might remain unrealized.

Third-Person Judgments

Third-person judgments present claims without anchoring them to any specific speaker or to the speaker's personal experience. Whether referring to named individuals ("Alice observed…"), using third-person pronouns ("She observed…"), or employing fully impersonal constructions ("It is observed that…"), this mode of expression deliberately withholds the narrator's presence from the act of knowing.

Consider these variations on a scientific claim about pendulums:

> Alice observed that all pendulums of length L in a vacuum near Earth's surface have period $T = 2\pi\sqrt{L/g}$.
>
> She observed that all pendulums of length L in a vacuum near Earth's surface have period $T = 2\pi\sqrt{L/g}$.
>
> It is observed that all pendulums of length L in a vacuum near the Earth's surface have period $T = 2\pi\sqrt{L/g}$.

In the first two cases, the speaker attributes accountability for the truth of the judgment to a particular individual—Alice or the person referred to by "she." The truth of the claim remains tied, however indirectly, to that specific observer (Alice or the "she") and her act of observation. The speaker, in making the statement, is merely reporting what the speaker has learned from Alice or the individual referred to by "she."

By contrast, the third case, a paradigmatic formulation in scientific discourse, presents the statement as if it belongs to the world itself, rather than to any particular speaker's perception or interpretation. The speaker recedes entirely into the background, allowing the claim to stand on its own as a fact, a finding, or a general law. By omitting first-person markers such as "I," the text avoids foregrounding any individual subject. "I observed *X*" becomes "*X* was observed," a reformulation that presents the result as observer-independent, *as though* it would hold true regardless of who made the observation, when or where it occurred, or whatever unmentioned contextual factors might be present. In the pendulum example, the judgment is framed as if it holds universally within the explicitly stated conditions (i.e., "in a vacuum near the Earth's surface"). The claim appears to transcend all other contingencies beyond those specified parameters: it seems indifferent to the observer's identity, the time or location of the experiment, or unstated variables such as ambient temperature, or the material composition of the apparatus. Third-person judgments, especially the fully impersonal "it-judgments" (e.g., it is observed that…), thus present or at least *aspire* to present a point of view that can, in principle, be adopted by *any rational subject*, irrespective of their specific perspectives or background conditions[3]. It is in this sense that a third-person judgment is *not* attached to *any specific subject*.

However, it must be emphasized that phrasing in a third-person form may function less as a mark of genuine objectivity than as *a rhetorical strategy* that *simulates* neutrality without necessarily achieving it. Framing a judgment in the third person can create the appearance of detachment, but this linguistic form alone offers no assurance that the underlying perspective has been purged of unexamined assumptions, implicit biases, or context-specific commitments. After all, even the most impersonal-sounding claim is still authored by someone situated within a particular historical, cultural, and institutional framework. The mere absence of first-person pronouns does not eliminate the influence of those conditions; it may simply render them invisible.

Moreover, the impression of context-independence conveyed by third-person formulations is often illusory. Even when speakers, whether individual researchers or entire scientific communities, sincerely aspire to neutrality and impersonality, their judgments may still subtly encode perspectives that remain unacknowledged: deeply ingrained cultural assumptions, anthropocentric biases, or limitations inherent to Earth-bound observation itself. The very instruments, categories, and standards through which people interpret the world are shaped by their historical and material conditions, yet these contingencies can disappear behind the polished surface of objective-sounding discourse. What presents itself as universally valid and detached from any particular standpoint may, upon closer scrutiny, reveal itself to be profoundly shaped by the very situatedness it claims to have transcended.

Relationships Among Different Types of Judgments

The various forms of judgment outlined above are not mutually exclusive; rather, they are deeply interwoven.

"I-Form" first-person judgments can be either judgments on direct or immediate experience ("direct judgments") or judgments derived from direct or immediate experience ("derived judgments"). An "I-Form" first-person direct judgment arises when the speaker reports an immediate, unmediated experience. In contrast, an "I-Form" first-person derived judgment results when the speaker draws an inference from their own prior experiences or observations, synthesizing past data into a new interpretive claim. Both remain anchored in the speaker's situated perspective, yet only the former claims empirical immediacy.

"We-form" first-person judgments, by contrast, are inherently indirect from the perspective of any single individual. No member of the "we" has direct access to the perceptual or experimental experiences of the others; instead, each treats fellow members' reports as testimonial evidence. Consequently, collective "we" statements necessarily involve interpretation, coordination, and epistemic trust. Even when such assertions appear seamless or authoritative, as in "We conclude that the hypothesis is supported," they do not rest on a unified perceptual field, but rather on a network of mediated inputs, negotiated understandings, and shared background assumptions.

Third-person judgments, by definition, do not express the direct experience of any single, identifiable subject. Instead, they report, synthesize, or abstract from the judgments of one or more observers. Consequently, all third-person judgments are indirect: they are necessarily derived, either explicitly or implicitly, from first-person accounts. These judgments may concern particular events ("It was observed that the sample degraded within 24 hours") or generalized regularities ("Water boils at 100°C at sea level"). Notably, for a third-person judgment employing fully impersonal constructions ("It is observed that…"), it aggregates multiple first-person observations into a shared, impersonal claim. Yet this very impersonality formulation obscures the human plurality and situated labor from which the judgment emerges.

Table 4.1 provides an overview of the distinguishing features of various judgment types and the relationships among them.

It bears emphasis that direct judgments are always tied to singular, concrete events. They are issued by specific subjects at determinate times and places,

Table 4.1 Relationship among different types of judgments

Feature	*"I-Form" first-person*	*"We-Form" first-person*	*Third-person*
Author	Single, identifiable subject ("I")	Plural, collaborative group ("we")	No specific speaker; impersonal
Epistemic basis	Own sensory or instrumental experience	Mix of self-experience and others' testimony	Aggregated, reported, or theoretical claims

through immediate sensory or instrumental engagement with the world, for example: "At 10:03 a.m., I recorded a pendulum swing of 2.01 s." Such judgments are inherently particular. In contrast, all generalized judgments are indirect, built upon a series of "I-form" reports of immediate experiences. Through comparison, abstraction, and pattern recognition, across multiple observers or across repeated trials by a single observer, these singular acts are reconfigured into broader claims about regularities, tendencies, or laws.

Publicity and "We/It" Form of Judgments

In Chapter 2, I treat publicity as that which transcends the private cognition of any single individual. A public judgment must be collectively accessible, capable of being recognized, assessed, and endorsed by multiple competent inquirers. As noted earlier, a first-person "we-form" judgment employs "we" to signal that the judgment could, in principle, be replicated or endorsed by multiple researchers, or even by any competent researcher working under similar conditions. It may also express an aspiration: that the judgment ought to be accepted by the scientific community once it has undergone proper scrutiny, even if full consensus has not yet been achieved. Similarly, third-person judgments, especially the fully impersonal "it-judgments," present, or at least aspire to present, a point of view that can, in principle, be adopted by any rational subject, irrespective of their specific perspectives or background conditions. Thus, both first-person plural ("we-form") and third-person impersonal ("it-form") judgments—the "we/it" forms—aspire to publicity and serve as vehicles for public knowledge. They are deliberately formulated to aspire to be intelligible, evaluable, and reproducible by others beyond their originator. In doing so, they transcend the privacy of individual experience and strive toward multi-personal validity. Conversely, it is precisely through the linguistic structures of these "we/it" forms that public knowledge must be articulated. Their very grammar enacts the conditions of publicity, making claims available not just to "me," but to "us," and even to "anyone."

Moreover, because publicity is a necessary condition for scientific objectivity, any scientist who wishes her judgments to count as objective therefore must formulate them in ways that aspire to collectively accessible, capable of being recognized, critically assessed, and potentially endorsed by multiple competent inquirers. This requires expressing those judgments either in a "we-form," which signals at least the possibility of multi-personal validity or shared endorsement, or in an "it-form," which aspires to articulate a standpoint that, in principle, any rational subject could adopt, regardless of their particular perspective, background, or circumstances. At a minimum, the "we-it" forms open the judgment to the public space of reason, where it may be tested, contested, and, if warranted, accepted as part of our shared epistemic repertoire.

> We know or we believe the model predicts a 2.7 K cosmic microwave background.
>
> It is believed that spacetime curvature explains gravitational lensing.

Thus, such statements in science are deliberately framed so that, given appropriate training and access to evidence, any qualified observer, regardless of origin, institutional affiliation, or prior belief, can assess their warrant. Their truth or falsity is not anchored in a single individual's mind but is open to evaluation by a community of inquirers. It is precisely this feature that *necessitates* their collective "we/it" formulation.

In this sense, scientific publicity is not merely a matter of availability, as if posting a paper online or shelving a book in a library were sufficient. Rather, it is a matter of *epistemic openness*: the *structural possibility* of critical engagement, replication, and rational assessment by a community of competent practitioners. It is this capacity of moving from the solitary "I" to the communal "we/it" that distinguishes objective knowledge from mere personal insight.

However, the shift from "I-form" judgments to "we/it" form judgments is not a mere stylistic convention of scientific discourse; it is a constitutive act of publicity, a necessary step in the very formation of scientific objectivity. This transformation is not merely rhetorical but epistemologically significant: it marks the passage from private experience to public knowledge. An "I-form" judgment originates in a concrete, embodied subject, situated in a definite physical and social context. It arises through the use of specific instruments, under particular background conditions (e.g., temperature, lighting, theoretical assumptions), and is filtered through the observer's unique perceptual and cognitive apparatus. Such a judgment speaks only for its author, articulating her personal understanding of a subject matter which often is an event she alone directly experienced. The speaker alone bears full epistemic *responsibility* for the claim, which remains tethered to her singular perspective and moment of observation. In contrast, "we/it" form judgments aim to be intelligible, evaluable, and reproducible by others beyond their originator. They seek to transcend the privacy of individual experience and aspire to multi-personal validity. Responsibility for such claims is distributed across multiple subjects, even across the entire scientific community, and embedded within shared norms of evidence, calibration, instrumentation, and reporting. The claim is no longer "mine," but potentially "ours" or even "any competent observer's." This divergence reveals a fundamental gap between "I-form" and collective "we/it" form judgments, reflecting the two core problems of publicity explored in Chapter 3. Especially, when an "I-form judgment" is a direct judgment—that is, a report of the author's immediate experience of a particular event arising from her interaction with a specific physical object or environment—its truth depends on two interrelated factors: how the *this* "I" interprets or conceptualizes that observation, and the concrete nature of the *interaction between this "I" and her immediate physical surroundings at that moment.* As discussed in Chapter 3, neither of these components is directly accessible to other subjects.

The first factor, how this "I" understands her observation, touches on the problem of other minds and connects to Davidson's account of objectivity: since we cannot directly access another's mental states, we rely on shared practices, language, and behavior to infer meaning. This constitutes *the interpersonal intelligibility dimension of publicity.*

The second factor, the particular physical interaction between the observer and her immediate environment, relates to *the referential identity dimension of publicity.* Even if others attempt to replicate the conditions, they cannot occupy the same spatiotemporal location or re-enact the exact material encounter. Consequently, they lack genuine, direct access to the very event that grounds the original judgment.

The gap between "I-form" and "we/it"-form judgments thus is not a matter of phrasing or stylistic preference; it is fundamentally epistemic. It poses a profound epistemological challenge: How is the transformation from "I" to "we/it" possible? At its core, this is the question of how a fragile, context-bound report of private experience can become a robust, public piece of scientific knowledge? How private, perspectival encounters are transmuted into multi-subjective valid claims? Or how can knowledge that originates in a singular, embodied perspective be rendered public without losing its empirical grounding? These are exactly the problem of publicity, reframed through the lens of the "I–we/it" gap.

Crucially, bridging the gap between "I-form" and "we/it-form" judgments cannot be achieved simply by having a subject "I" mechanically replace "I" with "we" or "it" in a judgment and then ask others to endorse the reformulated statement. Such a reformulation does not constitute genuine multi-subjective agreement; it merely expresses the speaker's *wish* that the claim be accepted by multiple subjects as if it reflected their perspectives. In reality, the reformulation remains entirely rooted in the "I's" own viewpoint and wish. It does not, by itself, embody or reflect the standpoints of others.

Even if the "others" asked by the "I" on the reformulated judgment by "I" agree with the reformulation, the agreement based solely on testimony does not amount to a genuine corroboration. A genuine corroboration requires *referential identity*: the judgment must concern the same subject matter for all involved observers. Only when different subjects are engaging with the same event, object, or interaction can their judgments be genuinely compared, tested, or collectively validated. Yet, as previously argued, this condition confronts a fundamental obstacle. When an "I-form" judgment reports a direct, immediate experience, the unique encounter between a specific observer and a particular physical situation, that observation is irreproducible in its full concreteness. The precise configuration of spatiotemporal context, instrument state, environmental conditions, and cognitive focus cannot be exactly re-enacted or directly accessed by others.

Consequently, asking others to agree with one's "I-form" direct judgment about such an episode is epistemically impossible: they lack direct access to the very event that grounds the claim. Any "agreement" they might offer on a reformulated version produced merely by mechanically replacing "I" with "we" or "it" does not constitute genuine multi-subjective or intersubjective endorsement. In other words, a "we/it-form" judgment generated through such superficial linguistic substitution fails to serve as a genuine public representation of a shared subject matter. It remains, at best, a private report dressed in collective syntax.

To bridge the epistemic gaps between private "I-form" judgment and public "we/it" judgment, I propose a three-step process:

Step 1: Step back from the egocentric "I," the concrete, situated observer, to a space of "possible subjects."
Step 2: Starting with "I-form" judgment, construct conjectural "we/it" judgment from the perspective of these possible subjects that reflects what aspire to be agreed upon under specified conditions.
Step 3: Test conjectural "we/it" judgment claim through communication or validation, thereby transforming conjectural "we/it" judgment into public "we/it" judgment.

In this approach, publicity and hence scientific objectivity is an achievement that requires deliberate epistemic labor.

Stepping Back From "I" to "Possible Subject"

From Impersonal "We/It" to "Possible Subject"

In previous discussions, the contrast between "I-form" judgments and "we/it" judgments has primarily centered on the fact that the former are fundamentally anchored in the epistemic standpoint of a real, situated "I." This "I" is not an abstract observer but a flesh-and-blood individual embedded in a specific physical and social context, equipped with particular tools, shaped by background conditions, and interpreting data through their own perceptual and cognitive apparatus. By contrast, "we/it" judgments are impersonal in form. They aspire either to multi-personal validity, expressing what "we" (a community of subjects) can jointly affirm, or to a universal characteristic of the "it-form," which presents a claim as adoptable in principle by any rational subject, regardless of their particular perspective or circumstances. Thus, judging from the differences in "I-form" judgment and "we-it" judgment, to construct a "we/it" judgment from an "I-form" judgment requires the original speaker, the specific "I", to step back from her singular standpoint and adopt a stance that is, in principle, shareable.

At first glance, "we/it" judgments appear neutral, objective, even voiceless. Statements like "The period of a simple pendulum is given by $T = 2\pi\sqrt{L/g}$" seem to float free of any speaker, as if uttered by nature itself. This stylistic convention fosters the illusion that science speaks without a voice or a view from nowhere. Yet this apparent detachment is just one way of understand "we/it" judgment. It can also be understood that a "we/it" judgment does not achieve authority by erasing subjectivity; rather, it succeeds by making a promise: that the claim it expresses could be affirmed by anyone among the multiple subjects or any subject under specified conditions. It invites others not merely to believe, but to verify for themselves. In doing so, it implicitly addresses and actively constructs a community of "*possible subjects*."

Consequently, "we/it" judgments should not—and indeed cannot—be understood as voiceless or neutral utterances floating free of any speaker. Rather, they are spoken on behalf of "*possible subjects*." These "possible subjects" are not actual individuals situated at a particular time and place, nor are they already embedded in the specific environmental conditions or background assumptions that shaped the original observation. Instead, they are *hypothetical yet concrete* agents: individuals who might enter the relevant experimental or observational setting, take up its instruments, engage with its material conditions, and participate in its prescribed interactions, provided they satisfy a set of minimal epistemic and practical criteria: access to appropriate tools, adherence to shared protocols or standardized procedures, basic theoretical literacy, and the capacity to follow established methodological norms. Phrases like "It was observed that…" or "Data show that…" are shorthand for: "Any possible qualified observer, operating under these defined conditions, would arrive at this conclusion."

This reframing carries significant implications. It reveals that the "we/it" is not the opposite of "I," but a reconfiguration of it. The authority of scientific claims does not stem from transcending human perspective, but from multiplying and structuring it. *The "I" is not eliminated; instead, it is projected outward into a space populated by possible others.* The resulting judgment thus carries within it a latent plurality: *it speaks not from nowhere, but from the standpoint of anyone who could be there*, a standpoint that is generic in form yet, in principle, inhabited by real and diverse individuals now and in the future.

In this sense, the impersonal form does not erase subjectivity but generalizes it, projecting the standpoint of the actual observer onto a space of epistemic possibility open to all competent participants. The impersonality of "we/it" judgment thus could be understood as "plurally personal." It trades the irreproducibility of a single concrete witness for the "reproducibility" of a shared epistemic stance of the "possible subject." And it is precisely this shift, from the unique "I" to the replaceable "anyone," that enables "we/it" judgment to circulate beyond its origin, and becoming public, not because it lacks a subject, but because it promises to *speak for many*: for anyone who, under the right conditions, could occupy the same observational or experimental position and arrive at the same conclusion.

It should be emphasized that the "possible subject" is not a metaphysical ideal—an abstract, uniform observer stripped of all particularity. On the contrary, it is a pragmatic and representational construct, deliberately designed to stand in for a plurality of actual, embodied subjects, each situated within their own distinct material, social, and cognitive circumstances.

Consider two observers, Alice and Bob, studying pendulum motion again. Alice conducts an experiment to test the standard pendulum law, and Bob subsequently attempts to replicate Alice's procedure as faithfully as possible. Yet, as argued earlier, exact replication is impossible. Subtle differences, in bob density and shape, thread elasticity, air currents, ambient temperature, timing resolution, or even how "release" is interpreted, inevitably introduce variation. These are not mere noise; they are constitutive features of any real-world experimental act. Nevertheless, if the outcomes of Alice's and Bob's respective "I-form" judgments based on their direct observation are sufficiently close to the relationship described by the "we/it" formulation of the pendulum law (e.g., $T = 2\pi\sqrt{L/g}$), then both may be *recognized as legitimate validations* of that same general "we/it" claim. However, their results will rarely, if ever, be numerically identical, but, within a shared framework of acceptable approximation and error tolerance, they count as convergent evidence. This inclusion is no accident. Scientific practice does not demand exact identity among observers and their background conditions; it accepts comparability under normatively defined conditions.

The "possible subject," therefore, does not represent a single ideal type but encodes a spectrum of real-world variability with differences in instrumental fidelity, environmental stability, theoretical background, perceptual acuity, etc. What matters is not that all observers are the same, but that their differences are *deemed* irrelevant relative to the specific claim at hand. Crucially, "we/it" judgments do not ignore this diversity; they *structure* it. A generalized statement such as "For small amplitudes, the period of a simple pendulum is independent of mass" carries implicit boundaries: it holds only when air resistance is negligible, the pivot is sufficiently frictionless, the amplitude remains below ~10°, and timing precision exceeds a certain threshold. Thus, the generalized statement implicates *deliberate methodological constraints* that carve out *a space of epistemic equivalence*, a zone within which diverse particular "I"s can nonetheless produce results that are mutually intelligible, intertranslatable, and, ultimately, collectively endorsable.

In this sense, the "we/it" judgment functions as a cartography of legitimate observational contexts—a structured space of possible subjects that delineates not where knowledge is, but where it can be reliably produced. Each point within this space corresponds to a real, situated observer: a specific person, at a particular time and place, working with available instruments, local calibration standards, institutional protocols, and background theoretical commitments.

The "possible subject" gains its representational power by mediating between the singular and the general. It allows a locally grounded judgment—one person's observation in a specific lab at a given time—to achieve wider validity, not by escaping context, but by being designed to resonate across contexts. Private experience becomes public knowledge not through uniformity, but by arranging differences within a common epistemic horizon where mutual recognition is possible.

This bridging function is essential to the social aspects of science. Without it, every "I-form" first-person judgment would remain isolated. The "possible subject" enables coordination across dispersed subjects, allowing them to build cumulative knowledge despite never sharing the same space and time, instrument, or even theoretical paradigm. It is not a denial of embodiment or situatedness, but a structured acknowledgment of their multiplicity, a way of saying: "Not just I, but others like me, under these conditions, would see the same." And in that promise lies the very foundation of scientific publicity.

"Possible Subject" as a Structured Projection of Variability

For the "possible subject" to function as a credible representative of diverse actual observers, it should not simply assume uniformity across contexts. Uniformity is neither empirically available nor epistemically desirable. Instead, the "possible subject" must actively accommodate and organize differences among real-world observers and their material conditions. This is where the epistemic labor becomes visible. Generalized claims are not mere extrapolations from repeated observations; they are carefully bounded statements that specify under what conditions certain variations are negligible and where differences become significant.

Consider again the pendulum example. The "we/it" judgment asserts that the formula $T = 2\pi\sqrt{L/g}$ holds regardless of *who* the observer is, *what mental state* she is in, or *which materials* she uses for the bob and string, whether steel or wood, silk or cotton, as well as her choice of timing device, measuring tool, or even release angle. This "we/it" judgment is thus not merely a truth about the world "as it is," but a deliberately constrained claim that defines the scope of *invariance across a range of possible subjects.*

A scientist could, if she chose, enumerate all parameters explicitly in her report: acceptable angle ranges, permissible materials, required timing precision, environmental tolerances, and so on. In doing so, she would be delineating *the space of admissible variation* within which diverse "I"s, including her own, count as legitimate *instances* of the "possible subjects." Each actual observer is a concrete point within this space; the generalized law holds *not* because all points are identical, but because they fall within a normatively defined zone of *equivalence.*

Thus, the power of the "we/it" form lies not in its detachment from context, but in its capacity to encode context selectively, to transform a multitude of embodied, situated acts of knowing into a coordinated field of potential agreement. The shift from "I" to "we/it" is therefore not a leap into publicity or objectivity, but a methodically constructed broadening of perspective—wherein the "possible subject" embodies the imprints of many or even innumerable situated observers.

The formula $T = 2\pi\sqrt{L/g}$ is not a direct judgment of any actual pendulum in the world. No real bob is a true point mass; no string is perfectly massless or rigid; and air resistance is never entirely absent. Rather, the formula functions as a theoretical anchor within a carefully delimited domain of applicability. Its validity is not absolute but relational: it holds to a good approximation when certain idealizing conditions are sufficiently met. These conditions, small angular displacement, a dense bob, a low-drag environment, are not incidental embellishments. They constitute the epistemic scaffolding that enables radically different experimental setups to produce commensurable results.

Again, a scientist could, if she chose, enumerate all these parameters explicitly in her report: for instance, specifying that "the observations were taken with amplitudes below 8°, using a brass bob of 500 g suspended by a nylon thread of 0.5 mm diameter, in still air at 22°C." Such details are far more than mere technical footnotes. They actively delineate the boundaries of *a shared observational space.* Each specification situates the experiment within a multidimensional parameter space: one dimension captures material properties (e.g., bob density, thread elasticity); another encompasses environmental conditions (e.g., temperature, air currents); a third reflects the particular mental states and procedural choices of the observing subject—the specific "I"—including release angle, timing method, and the instruments used (clock, ruler, etc.); and a fourth encodes the unique event itself—the dynamic interaction between that subject and her immediate experimental environment, and so on. Within this bounded region, differences among individual "I"s across laboratories, instruments, or even theoretical orientations are rendered epistemically inert. Outside it, however, those same differences may become decisive, undermining comparability and casting doubt on the validity of the generalization.

Thus, the "possible subject" is far from a neutral placeholder. It is a *structured projection* of *mental and physical variability*. Thus, a shift from the first-person "I observed a period of 2.01 s" to the third-person "It is observed that the period follows $T = 2\pi\sqrt{L/g}$" is neither a magical leap into publicity nor a denial of perspective. Rather, it is a carefully scaffolded expansion of perspective, achieved through practices that align multiple viewpoints in epistemically relevant ways.

In this light, scientific publicity and, of course, scientific objectivity, if we retain the term at all, must be reconceived not as freedom from subjectivity, but as robust multi-subjectivity under specified conditions. The "possible subject" embodies this ideal not by transcending difference, but by orchestrating it, making room for diversity while securing coherence. It is this delicate balance between openness to variation and demand for consistency that enables scientific knowledge to remain rooted in practice by concrete, imbodied subject, while retaining the capacity to travel across various real individuals in laboratories, disciplines, cultures, and generations. It is precisely because we acknowledge mental and physical variability that scientific practice has developed evolving methodological norms to navigate it. Through historical negotiation, empirical testing, and communal consensus, these norms determine which aspects of situatedness are epistemically relevant and which may be safely bracketed as negligible. What gets set aside, say, minor temperature fluctuations, and what must be rigorously controlled, such as release amplitude, reflects, on one hand, deep epistemic judgments about the stability and robustness of the phenomenon under investigation, and, on the other hand, an attitude on possible variable parameters or possible situations.

From this perspective, traditional ideals of scientific objectivity, such as value detachment and disinterestedness, are not just attempts to deny or ignore the pervasive role of values and interests in scientific practice. On the contrary, they represent a sophisticated acknowledgment that values and interests are constitutive features of human cognition and, more specifically, intrinsic dimensions of mental and epistemic variability. These norms did not arise from an unrealistic belief that scientists could or should ever become entirely free of values or interests. Rather, they emerged from a historically grounded insight: that unexamined or unchecked idiosyncrasies, whether rooted in personal bias, institutional pressures, disciplinary orthodoxy, or deep-seated theoretical allegiances, can introduce destabilizing forms of variation into observation and reasoning. Such variation risks trapping first-person ("I") reports and judgments within the confines of a single mind, thereby obstructing the development of a multi-subjective foundation essential for publicly credible ("we/it") knowledge claims. Consequently, value detachment and disinterestedness may not be interpreted as efforts to eliminate values, interests, or subjectivity altogether. Instead, they function as disciplined, reflexive strategies for managing these very elements. They embody a methodological stance that encourages scientists to step back from the immediacy of their own situated "I" with its particular values and interests and orient themselves toward the "possible subjects": a horizon of generic, interchangeable observers who may not share the same commitments, yet, under shared methodological conditions and communal epistemic norms, can converge on mutually acceptable conclusions for the sake of public accountability and communicability.

On one hand, this orientation is operationalized through collective epistemic practices that specify when, how, and to what extent certain situated influences, such as personal values or interests, might affect or even "distort" scientific judgment. On the other hand, these same practices provide mechanisms by which such influences may be *provisionally* bracketed, neutralized, balanced against competing values, or otherwise held in check. Crucially, this does not entail that such influences should be entirely denied or eliminated. Values may indeed remain necessary at certain stages of scientific practice, for instance, in evaluating the societal risks of specific research programs. Rather, the point is that when values are legitimately required, scientists can reflexively attend to the relevant subjects of evaluation and deliberately constitute them as value-oriented agents, that is, as participants whose judgments are explicitly guided by ethically or socially informed considerations, rather than by unexamined bias.

Such regulation becomes especially important when researchers who hold divergent commitments seek at least a minimal degree of common ground—an essential precondition for collaboration, which often requires mutual compromise. Thus, far from aiming to purge science of its irreducibly human dimensions, these norms seek to channel them constructively: not to eliminate difference, but to coordinate it. In doing so, they enhance not only the robustness, accountability, and cross-subjective validity of scientific knowledge, but also the practical capacity of diverse researchers to work together effectively within shared scientific enterprises.

Procedural objectivity, (Megill 1994)[4], by contrast, does not seek to eliminate human or material variability altogether. Rooted in standardized protocols, calibrated instruments, and repeatable methods, it functions not as a barrier against difference but as a selective filter—a methodological sieve that sorts variation into two epistemic categories. On one side are epistemically disfavored forms of variation: idiosyncratic influences that imprint private, contingent features onto scientific results, such as individual judgment calls, unrecorded assumptions, or ad hoc adjustments. These threaten the reproducibility and public intelligibility of findings by tethering knowledge too closely to the particularities of a single observer or context. On the other side are epistemically favored forms of variation: differences that either reflect genuine features of the phenomenon under study or can be systematically accounted for within a shared framework. These variations are not suppressed but preserved, as potential signals rather than noise.

By codifying precise procedures for data collection, instrument operation, environmental control, calibration, and error reporting, procedural objectivity constructs a normative scaffold that enables any competent observer to act as if they were a "possible subject" who is an attainable stand-in for the broader scientific community, committed not to personal insight but to publicly accessible results. Within this framework, sources of unwanted difference, such as observer bias, inconsistent timing, fluctuating ambient conditions, or undocumented methodological drift, are not ignored but actively neutralized, minimized, or rendered transparent through protocol. Crucially, procedural objectivity does not render science impersonal by stripping away context; rather, it orchestrates context. It deliberately designates certain dimensions of variation, not as idiosyncratic to a single observer, but as shared across a broad range of, or even all possible, subjects, while treating other dimensions as specific to particular observers. This distinction between "signal" and "noise" is not imposed by arbitrary fiat; rather, it emerges from historically accumulated, communally tested, and continually refined practices of epistemic discernment—practices fundamentally oriented toward public coordination and scientific cooperation. In this sense, procedural objectivity is less about achieving value-free neutrality and more about establishing a shared grammar of publicity—one that allows diverse researchers to coordinate their efforts, compare results across space and time, and build cumulative knowledge despite their inevitable differences in background, perspective, or interest.

Constructing Conjectural "We/It" Judgment

A first-person "I-form" judgment only speaks for one specific subject. A "we/it" judgment speaks on behalf of *"possible subjects" which represent many subjects other than some specific I.* To construct a "we/it" form judgment from "I-form" direct judgment requires this "I" to first step back from this egocentric "I" and adopt a stance that is, in principle, shareable among multi-subjects or even all competent subjects as explained earlier. However, if a "possible subject" is understood as a structured projection of mental and physical variability, how could this structure itself be constructed?

Empathetic Representation

A possible answer to how the "possible subject" is structured lies in the scientist's active, imaginative, or anticipated engagement with variability, not as a passive recipient of data, but as a prospective modeler of alternative experimental enactments. This modeling is not limited to other human observers; it extends empathetically to materials, instruments, and environments, anticipating how they might behave under altered or counterfactual conditions. Scientific practice, in this view, does not begin only at the moment of data recording. It commences earlier in the mind where the researcher, as a situated "I," mentally rehearses variations of the experiment before a single measurement is taken.

She might ask: What if this procedure were carried out by a colleague in another lab using a slightly older photogate? What if the pendulum bob were cast in brass instead of steel? What if the entire experiment were repeated on a hypothetical day when, globally, all clocks ticked half as fast and all rulers shrank to half their length?

These are not idle speculations or mere flights of fancy. They are disciplined acts of epistemic imagination, grounded in the scientist's empirical history, her awareness of experimental fragility, and her immersion in the collective memory of the scientific community. Such simulations draw on hard-won lessons from failed replications, contested calibrations, methodological controversies, and even carefully constructed thought experiments. Through this process of anticipatory reasoning, the scientist probes the robustness of her conclusions, testing how they hold up across a landscape of plausible variations, both human (differences in training, interpretation, or instrumentation) and material (changes in substance, environment, or other physical parameters).

In doing so, she constructs the "possible subject" as a dynamic constellation of imagined others and alternative conditions. Moreover, this anticipatory work enables her to pre-emptively embed variability into the very architecture of her epistemic claim. Rather than anchoring her assertion solely in a private, first-person "I-form" observation ("I saw…"), or *merely replacing* the "I" with "we/it" *in hopes of* securing agreement or endorsement from others, she *constructs* a *conjectural* "we/it" judgment: a formulation such as "It was found that…" or "The data indicate…," which speaks not only for herself but for a "possible subject." This "possible subject" is a carefully delineated epistemic position that *any* competent researcher (including the original observer under different circumstances) could occupy, provided they adhere to specified methodological conditions and account explicitly for relevant human and material variabilities.

Crucially, this move is both strategic and epistemically generative. Rather than *passively awaiting* the emergence of discrepancies across contexts, the scientist *proactively designs* her claim to already accommodate a bounded space of legitimate variation in training, instrumentation, timing, environmental conditions, and even interpretive dispositions. The resulting "we/it" statement is thus not a mere record of a singular occurrence, but *a conjectural assertion*, deliberately crafted to hold across a spectrum of possible enactments, provided that core methodological constraints are upheld.

Importantly, the conjectural "we/it" form judgment is neither arbitrary nor idealized. Its capacity to circulate across diverse contexts stems precisely from its responsiveness to anticipated variability that shaped not only by the collective history of scientific practice but also by the personal, dialogical history between the "I" who articulates the judgment and the broader community of researchers. Embedded within such a judgment is a latent map of empathetically calibrated conditions of validity: a tacit yet structured framework specifying which parameters matter, how their acceptable ranges are determined, and which sources of variation may be provisionally set aside without jeopardizing the conclusion. Even the act of declaring certain factors insignificant, such as minor fluctuations in ambient temperature or slight differences in observer reaction time, is not a dismissal of context, but a deliberate engagement with it. This selectivity reflects not oversight, but an epistemic choice grounded in empirical experience, theoretical insight, and historical precedent.

In this light, the "possible subject" is neither a metaphysical fiction nor a statistical average. It is a pragmatic construct, actively forged through practices that anticipate difference, not to eliminate it, but to channel it into a shared framework of mutual intelligibility. Scientific publicity, then, does not require the erasure of the situated "I." Rather, it arises through the orchestration of multiple perspectives into a coherent, transportable form of knowledge—one that can be inhabited, verified, critiqued, and extended by any suitably positioned observer, now or in the future.

Variability and Invariance

The anticipatory work undertaken by a specific "I" in constructing a possible subject requires her to "preemptively" embed variability into the very architecture of her epistemic claim. She does not merely report what she saw; she formulates a conjectural "we/it" judgment that speaks not only for herself but for an anonymous, possible subject who might occupy the same epistemic position under specified conditions. This includes the original observer herself, with relevant human and material variabilities explicitly taken into account.

This brings the specific "I" of the "I-form" judgment to a core requirement of any "structured projection of mental and physical variability": invariance. For a conjectural "we/it" judgment to *qualify* as potentially public, and thus be potentially endorsable by others, it must remain stable under transformations across the space of possible observational standpoints. That is, regardless of which anonymous "I" occupies the position—whether Alice, Bob, in Germany, or in France, or a future Mars colony, the core claim must hold.

Also, consider the pendulum: individual measurements of its period may vary slightly due to differences in timing devices, human reaction time, or local temperature anomalies. Yet the conjectural "we/it" form relationship "$T = 2\pi\sqrt{L/g}$" must potentially be recoverable by *any competent observers*. Similarly, the judgment remains stable under minor variations in background conditions, such as small fluctuations in ambient temperature or humidity, provided these fall within the range of variability, which has been built into the "possible subject" or "we/it" judgment. Therefore, declaring certain variations irrelevant (e.g., "environment temperature does not affect the result beyond the stated error margin") is itself a form of disciplined engagement with variability, not its denial.

Importantly, invariance does not require identity of raw experience. Two observers may differ in sensory acuity, instrument precision, theoretical background, or even cultural framing. Yet if the conjectural "we/it" judgment is constructed through an appropriate "structured projection of mental and physical variability" that anticipates and accommodates legitimate sources of difference, it can remain robust across these variations. Invariance operates not at the level of sensation or private perception but at the level of structured, norm-governed representation: a shared epistemic framework in which diverse perspectives are coordinated.

In sum, the construction of a conjectural "we/it" judgment rests on two interdependent operations: empathetic representation, the transformation of a specific, embodied "I" into a possible subject, with mental and physical variability explicitly built into the epistemic claim; and invariance, the assurance that this conjectured knowledge for "possible subject" remains stable across the full ensemble of possible subjects and material conditions—thereby enabling first-person "I-form" judgments, though irreducibly private in origin, to support mutually translatable and collectively endorsable conclusions.

Publicity, therefore, is not the absence of perspective but the achievement of perspectival coherence. It is forged in the liminal space between the concrete and the possible, where the traces of countless actual "I"s are woven into a shared fabric of reliable, transportable knowledge. The possible subject is not a metaphysical fiction; it is the architectural scaffold that allows science to speak not just for one, but for all who might look, measure, and reason in the right way.

A Symbolic Formulation

The structure of the conjectural "we/it" judgment described above may be rendered more precisely through a formal symbolic representation.

In scientific practice, the "I" of an "I-form" direct judgment always refers to a specific, embodied subject: a real person at a determinate time and place, engaging with the world through particular instruments under concrete environmental, material, and cognitive conditions. Such an "I-form" judgment is not merely a report of raw data; it is the trace of a singular event, inseparable from its full constellation of circumstances, the observer's posture, the calibration history of the instrument, ambient temperature fluctuations, even momentary shifts in attention or expectation.

The speaker here is not a disembodied intellect but a flesh-and-blood agent whose perceptual engagement is inherently private. The truth of this "I-form" judgment is not determined by any single factor in isolation, neither by the mind of the observer alone, nor by the external environment alone, but by the entire configuration of interrelated elements: the observer's cognitive and bodily state, the immediate physical surroundings, the background conditions, and the specific interaction between them. In other words, the truth value of an "I" judgment is co-determined by a network of variables, all of which jointly constitute the epistemic situation in which the judgment arises.

Thus, an "I-form" direct judgment may be formalized as:

$$f^{\text{“}I\text{”}}\left(S^{\text{“}I\text{”}}, O, \{C_1, C_2, \ldots, C_m, \ldots\}, E\right),$$

where $f^{\text{“}I\text{”}}$ is the "I-form" judgment by *this* specific subject "*I*," $S^{\text{“}I\text{”}}$ is the specific subject "*I*," O is the object or physical environment immediately observed by *this* "*I*," and $\{C_1, C_2, \ldots C_m, \ldots\}$ is the set of background conditions (e.g., air flow, various fields) which might affect the truth of the "*I*" judgment. E represents the event of interaction between this subject "*I*" and her immediate environment. The judgment's truth value emerges from the *entire configuration*, not from any single parameter in isolation.

A "we/it" reformulation of this judgment, constructed by the same subject "I" through anticipating variability, can be expressed as:

$$f^{\text{“}we/it\text{”}}\left(S^{\text{“}we/it\text{”}}, O^{\text{“}we/it\text{”}}, \{C_1, C_2, \ldots, C_m, \ldots\}^{\text{“}we/it\text{”}}, E^{\text{“}we/it\text{”}}\right),$$

where $S^{\text{“}we/it\text{”}}$ is a "possible subject" constructed by $S^{\text{“}I\text{”}}$, but with anticipated variations among observers built in (e.g., differences in perceptual acuity, cognitive framing, conventions of measurement such as clock rates or ruler lengths); $O^{\text{“}we/it\text{”}}$ is the observed object under possible material variations (e.g., temperature, composition, or state changes that fall within the domain of experimental relevance); $\{C_1, C_2, \ldots, C_m, \ldots\}^{\text{“}we/it\text{”}}$ includes background conditions extended to encompass plausible ranges of variation; $E^{\text{“}we/it\text{”}}$ captures whatever variations in interaction event, accounting for how different observers might engage with the object under varying conditions.

Hence, by constructing this "we/it" judgment, the original subject "I" introduces an additional layer of structure into the original tuple ($S^{\text{“}I\text{”}}$, O, $\{C_1, C_2, \ldots, C_m, \ldots\}$, E)—*a structured projection of mental and physical variability.*

Now, if we bracket the problem of referential identity—that is, if we set aside the irreproducibility of first-person immediate experience and focus solely on the Davidson's problem of objectivity (i.e., how shared content can arise across distinct minds)—then the "we/it" judgment may take a simplified form:

$$f^{\text{“}we/it\text{”}}\left(S^{\text{“}we/it\text{”}}, O, \{C_1, C_2, \ldots, C_m, \ldots\}, E\right),$$

In this version, the object O, background conditions $\{C_1, C_2, \ldots, C_m, \ldots\}$, and interaction event E are held fixed as they appeared in the original "I-form" judgment. Only the subject is generalized: $S^{\text{"we/it"}}$ now represents the original "I" imagining what other competent observers would experience if placed in her exact situation. This act enlarges her perspective through empathetic projection: she simulates how differences in attention, convention, or sensory capacity might play out, even while assuming identical external conditions—a shared stimulus. Even in this restricted case, the "I" embeds *a structured projection of human variability*[5] into her own standpoint.

Consider a transformation that replaces the specific "I" who constructs a "we/it" judgment with her own "I-form" judgment with one instantiation, other than this "I," of the "possible subject":

$$\left(S^{\text{"I"}}, O^{\text{"I"}}, \{C_i\}^{\text{"I"}}, E^{\text{"I"}}\right) \rightarrow \left(S_\mu^{\text{"we/it"}}, O_\mu^{\text{"we/it"}}, \{C_i\}_\mu^{\text{"we/it"}}, E_\mu^{\text{"we/it"}}\right)$$

This represents a shift from this specific I (for the convenience, I add a subscript to this specific subject "I"'s parameters in the tuple) to another possible subject (indexed by μ) who embodies distinct but admissible configurations of human and material parameters.

If the judgment function f remains invariant under this transformation—that is, if

$$f^{\text{"I"}}\left(S^{\text{"I"}}, O^{\text{"I"}}, \{C_i\}^{\text{"I"}}, E^{\text{"I"}}\right) = f^{\text{"we/it"}}\left(S_\mu^{\text{"we/it"}}, O_\mu^{\text{"we/it"}}, \{C_i\}_\mu^{\text{"we/it"}}, E_\mu^{\text{"we/it"}}\right),$$

then the resulting conjectural "we/it" claim constructed by the specific "I" upon her original "I-form" judgment has transcended private cognition. It is no longer tethered to a single embodied perspective, but is potentially endorsable by multi-subjects.

Moreover, consider a transformation that replaces an arbitrary instantiation of the "possible subject" with another arbitrary one:

$$\left(S_\mu^{\text{"we/it"}}, O_\mu^{\text{"we/it"}}, \{C_i\}_\mu^{\text{"we/it"}}, E_\mu^{\text{"we/it"}}\right) \rightarrow \left(S_v^{\text{"we/it"}}, O_v^{\text{"we/it"}}, \{C_i\}_v^{\text{"we/it"}}, E_v^{\text{"we/it"}}\right)$$

where $\mu \neq v$. This represents a shift from an arbitrary possible subject (indexed by μ) to another arbitrary possible subject (indexed by v), each embodying distinct but admissible configurations of human and material parameters.

If the judgment function f remains invariant under this transformation—that is, if

$$f^{\text{"we/it"}}\left(S_\mu^{\text{"we/it"}}, O_\mu^{\text{"we/it"}}, \{C_i\}_\mu^{\text{"we/it"}}, E_\mu^{\text{"we/it"}}\right)$$
$$= f^{\text{"we/it"}}\left(S_\mu^{\text{"we/it"}}, O_\mu^{\text{"we/it"}}, \{C_i\}_\mu^{\text{"we/it"}}, E_\mu^{\text{"we/it"}}\right),$$

then the resulting "we/it" claim is no longer tethered to a single embodied perspective; it is potentially endorsable by any single competent observer.

In this sense, publicity is achieved not through the elimination of perspective, but through invariance across a space of possible perspectives. The "possible subject" is thus not a metaphysical ideal, but a pragmatic scaffold—a structured template that enables singular experiences to become collectively intelligible.

Publicity Realization

The construction of publicity previously developed begins not with consensus but with a single, situated subject "I." This "I" must first step back from her egocentric standpoint and enter a conceptual space populated not by actual others, but by "possible subjects"—normatively defined observers who might occupy the same epistemic position under specified conditions. From this vantage, she constructs a conjectural "we/it" judgment: a "generalized" claim that aspires to be intelligible, evaluable, and potentially endorsable by others.

This act of "generalization" is not a neutral abstraction. This empathetic representation is a disciplined, two-way imaginative effort that anticipates variability across human observers. On one hand, it is an outward projection. The "I" imagines how other observers might differ from her in relevant respects, their mental states (e.g., attentional focus, theoretical commitments), their perceptual dispositions, or their practical conventions (e.g., choice of units, calibration standards, or instrument types). On the other hand, it is an inward counterfactual. Simultaneously, the "I" asks: What if another competent observer were placed in my exact situation? How might differences in their cognitive framing or material setup affect the judgment, even when observing the same phenomenon under seemingly identical conditions?

Thus, imagination, anticipation, or empathy here is not mere sympathy; it is a methodological exercise in perspectival simulation. The "I" does not only step into others' shoes, imagining how colleagues, instruments, or environments might respond under altered conditions, but also "invites" others into her own epistemic position, testing the robustness of her judgment against plausible variations in both human and material parameters.

It is through this dual-lens construction of the possible subject and the conjectural "we/it" judgment that publicity is brought to vision. But crucially, this publicity remains conjectural, a hypothesis about multi-subjective stability, not its realization.

To actualize publicity—to transform a conjectural claim into genuinely shared knowledge—a final, indispensable step is required: empirical validation through communication with actual others. However meticulously a conjectural "we/it" judgment has been crafted to anticipate variability, it remains provisional until it is tested against the judgments of real observers working in concrete laboratories, under real-world constraints, and within their own situated practices. This validation often begins with direct comparison to "I-form" reports from diverse observers—accounts that instantiate the abstract "we/it" judgment in relation to specific, embodied "I" perspectives. Though rooted in individual experience and bearing the idiosyncrasies of specific contexts, these reports serve as crucial anchors for cross-verification. Because the original "we/it" judgment already embedded a structured projection of mental and physical variability, successful alignment with others' judgments, even when those others differ in subtle ways, confirms that the core claim is invariant under legitimate transformations across the space of possible observers. Only at this point can the "I" justifiably declare: My conjectural "we/it" judgment has transcended my private cognition. It is no longer merely mine but is realized public knowledge. The following is an example for this third step of bridging "I-form" judgment and "we/it" judgment.

Consider again the pendulum example: an observer in China measures the swing time of a brass bob suspended by a nylon thread under a specific set of conditions—room temperature ≈ 20°C, minimal air currents, using a digital stopwatch she calibrated against her phone's atomic-clock-synced timer, etc. Her "I-form" judgment is: "I observed five full oscillations of this brass pendulum in 10.05 seconds; thus, $T = 2.01$ s."

Anticipating variability, she constructs a conjectural "we/it" judgment:

> For *any* pendulum of length $L = 1.00$ m, the period is approximately 2.01 s, assuming small-angle release, standard gravitational acceleration, and timing methods with ± 0.02 s precision.

In doing so, she builds in possible differences: others might use a steel pendulum instead of a brass pendulum, or photogates instead of stopwatches, measure at slightly different amplitudes, or work in labs with varying humidity or altitude (affecting *g* marginally).

She then shares her protocol and result with colleagues in Germany and France. The German observer uses a steel bob system; the French colleague works in a colder room (8°C) and counts swings manually but averages ten trials. Despite these differences in instrument choice, environmental conditions, and attentional focus, all three obtain periods between 2.00 and 2.02 s, consistent within expected error bounds.

Because her original "we/it" judgment already incorporated plausible mental and physical variability, and because the core claim remains stable across actual observers, the initial private inference is now validated as public knowledge: not because everyone saw exactly the same thing, but because their diverse yet legitimate practices converged on a shared conclusion.

It should be emphasized that this multi-personal process is not a one-time verification but an enduring, iterative practice—a continuous cycle of testing, feedback, and revision. Through this sustained intersubjective confrontation, the original conjecture is rarely simply confirmed or rejected in binary fashion. More often, it is refined, recalibrated, or even fundamentally reconfigured in light of new evidence, alternative interpretations, or unexpected experimental outcomes.

Discrepancies that emerge across replications or cross-laboratory comparisons are not merely obstacles; they are epistemically generative. They may expose unanticipated sources of variation, such as faulty assumptions about instrument stability, overlooked sensitivities to ambient conditions, ambiguities in procedural protocols, or tacit differences in theoretical framing, that were absent from the initial validity map embedded in the "we/it" judgment. Each such revelation prompts a re-examination of what counts as relevant variability and how robustly the claim can accommodate it.

Conversely, when diverse observers, working with different tools, in different locations, and under varying conditions, arrive at convergent conclusions, the epistemic standing of the conjectured "we/it" judgment is strengthened. Yet this convergence does not prove the conjectured "we/it" judgment's universality or absolute truth. Rather, it demonstrates its resilience within an expanding envelope of legitimate variation: the claim holds not because the world is uniform, but because the judgment has been successfully designed to remain stable across a widening range of plausible human and material configurations.

Thus, actualized publicity is not a static endpoint but an ongoing, dialogical achievement—a fragile yet durable alignment of perspectives continually forged through acts of communication, critique, replication, reinterpretation, and mutual adjustment. It is in this crucible of real-world engagement that the "possible subject" actually transcends its origins as a solitary act of empathetic imagination and becomes a living node in a distributed network of scientific practice: a shared reference point around which communities coordinate their inquiries, calibrate their instruments, and build cumulative knowledge. In this sense, publicity and, of course, objectivity is not discovered once and for all, but is sustained through collective vigilance and collaborative care.

Notes

1 Einstein, Albert. "A Generalized Theory of Gravitation." *Review of Modern Physics*, 20 (January, 1948): 35–39.

2 "We" may also gesture toward the ideal of objective knowledge, depending, of course, on how the term "objective" is understood. However, as previous discussions have shown, conceptions of objectivity are deeply contested and entangled. Therefore, I would prefer to set aside the term "objective" in this analysis of judgmental distinctions until a more appropriate moment arises.

3 Similarly, Pauen (2012) argued that the third-person perspective is uniquely unrestricted: while the first-person perspective is confined to the epistemic subject herself, and the second-person applies specifically to other subjects, the third-person perspective can be applied to virtually any object or phenomenon, including other minds, physical systems, or abstract relations. Pauen further emphasizes that this generality is what underwrites the perceived objectivity of the third-person stance. See Pauen, Michael. "The second-person perspective." *Inquiry*, 55 (February, 2012): 33–49.

4 Megill, Allan. "Four senses of objectivity." In *Rethinking Objectivity*. Edited by Allan Megill, 1–29. Durham: Duke University Press, 1994.

5 Human variability may encompass both mental variability (e.g., cognitive states, attentional focus, or interpretive frameworks) and physical variability, such as conventions in selecting or calibrating instruments.

5 Publicity Condition in Spacetime Dimension

In this chapter, I present a case study on the constitution of publicity in scientific theories, focusing specifically on the domain of spacetime structure. The analysis centers on Hermann Weyl's first unified field theory of 1918, which sought to unify Einstein's general theory of relativity—a geometric theory of spacetime—with electrodynamics. Although Weyl's proposal was ultimately ruled out by empirical evidence (notably Einstein's objection to atomic spectra) and is often cited as a historically significant but failed physical theory, I argue that its symmetry requirement can be interpreted as an implicit attempt to establish a condition of publicity[1] in the representation of spacetime.

Importantly, Weyl himself did not explicitly frame his symmetry principle in terms of publicity or intersubjectivity; his philosophical remarks on the matter remain vague and underdeveloped. Whether he intended his theory as a response to the problem of intersubjectivity is not my concern. Instead, I adopt a reconstructive stance on the problem of intersubjectivity in Weyl's theory. Here, I treat intersubjectivity in a broad, functional sense, namely, as the condition that objects and judgments be accessible to, and acceptable by, all competent observers. On this reading, intersubjectivity represents a stronger demand than publicity, which requires only that knowledge be accessible to multiple subjects (but not necessarily to all). Nevertheless, publicity remains a necessary condition for intersubjectivity: if a claim cannot be shared among even multiple observers, it certainly cannot be universally endorsable.

For the sake of coherence with the historical and conceptual details of Weyl's theory, I will mostly use the term "intersubjectivity" throughout the analysis of this case. Only when necessary and at the end of the chapter will I return to clarify whether Weyl's symmetry requirement should be understood not merely as a mathematical constraint, but as a constitutive condition for publicity, that is, as a formal mechanism designed to ensure that spacetime measurements transcend private experience and become part of a shared, public framework.

For the sake of clarity, and for the record, I should note that my engagement with Weyl's ideas in this chapter is not strictly exegetical. In several

DOI: 10.4324/9781003783787-6

places, I interpret or extend his concepts through the lens of the conceptual framework developed in previous chapters. This approach inevitably carries the risk of attributing to Weyl viewpoints or emphases that may not be authentically his. However, this is by design. My aim here is not to reconstruct Weyl's philosophy as it was, but rather to treat his so-called "failed" theory, not as a historical endpoint, but as a productive context for a philosophical case study. Specifically, I use it to illustrate and test the explanatory power of the theoretical apparatus I have articulated in preceding chapters. Consequently, even if certain interpretations diverge from Weyl's original intentions, such departures do not undermine my central purpose.

A Transcendental Idealism Spacetime Theory and the Problem of Intersubjectivity

Weyl's 1918 unified field theory and its underlying geometric framework, which he called "purely infinitesimal geometry," have been examined from historical and mathematical perspectives (e.g., Goenner 2004; O'Raifeartaigh and Straumann 2000; Ryckman 2020)[2]. Yet equally significant is the philosophical context in which Weyl situated his work. Deeply influenced by transcendental philosophy, particularly, though not exclusively, Husserlian phenomenology[3], Weyl sought to ground physical theory in an epistemology that begins with the immediacy of conscious experience.

Central to Weyl's project is his characterization of the coordinate system as "the unavoidable residue of the ego's annihilation" (Weyl 1994, 94)[4]. This striking phrase captures his ambition to bridge what he saw as a fundamental chasm: that between the inexact, intuitive continuum of lived experience and the exact, mathematical continuum of symbolic representations in physics. Researchers have rightly emphasized this tension as a core motivation behind Weyl's symbolic construction of spacetime (Ryckman 2005; Tieszen 2000)[5].

However, beyond this well-discussed epistemological divide, Weyl's schema confronts a second, equally profound challenge: the problem of intersubjective constitution. If objective knowledge is to be more than a private mental construct, it must be shareable or accessible not only to the originating subject but to others as well. Yet the "intuitively given," Weyl's epistemic starting point, is inherently subjective or private: it is directly available only to the individual consciousness that experiences it. As Weyl himself puts it, from a "purely epistemological point of view," we begin only with "an image, a vision, a phenomenon of our consciousness." The contents of this immediate experience—vague, qualitative, and inexact—serve as the raw material for rational reconstruction, driven by what Weyl calls "rational motives" to achieve a representation of the world that is "truly objective, exact, and non-qualitative" (Weyl 1994, 93)[6].

The introduction of a coordinate system functions, in this light, as a symbolic scaffold that lifts people from the flux of intuition toward mathematical precision. But here a new difficulty emerges: how can such a reconstruction be made intersubjectively valid? If my intuitive givens are mine alone and are private, how can others access, verify, or co-constitute the resulting physical claims?

Weyl acknowledges this dilemma. In *Philosophy of Mathematics and Natural Science* (1927), he writes:

> …if I furthermore accept as data the contents of the consciousness of others on equal terms with my own, thus *opening myself to the mystery of the intersubjective co*mmunication…
>
> (Weyl 1949, 117)[7]

Yet, unlike his meticulous treatment of the intuition–mathematics gap, Weyl offers no systematic account of how intersubjective communication is possible, or how shared objectivity is constituted from disparate subjective origins. His recognition of the problem remains gestural, almost aporetic: he identifies the "mystery" but does not resolve it.

It is precisely at this juncture that I argue the significance of Weyl's symmetry principle—not merely as a physical postulate, but as a proto-constitutive condition for intersubjectivity or publicity. Within Weyl's transcendental-idealist framework, intersubjective constitution must be understood as the process through which isolated subjects arrive at a common representational space in which judgments about spacetime properties can be mutually assessed and endorsed. While Weyl does not explicitly frame his gauge symmetry in these terms, I argue that the requirement of invariance under local scale transformations can be interpreted as an attempt, however implicit, to neutralize the residue of the ego and thereby enable a form of representation that transcends any single subject's perspective.

In short, if the coordinate system marks the "residue of the ego's annihilation" in the sense of from the inexact, immediate experience to exact symbolic representations, then the symmetry condition may be seen as the mechanism of that annihilation—a formal constraint designed to ensure that the resulting physical laws are not tethered to the convention of one observer's convention, but are instead accessible, in principle, to all.

Weyl's Purely Infinitesimal Schema

The Starting Points and Principles of Combinations

To have a secure foundation for the mathematical–physical mode of scientific cognition, Hermann Weyl insists that we must begin not with "more or less arbitrarily axiomatized systems" (Weyl 1994, 24)[8], but with what he calls the intuitively given which he designates both as "the starting point" and "the ultimate foundation" of scientific thought. In phenomenological terms, the immediately given serves a dual role: it is both the origin from which transcendent objects are symbolically constituted and the terminus toward which phenomenological reduction leads. It functions, in Weyl's words, as *an Archimedean point*, whose meaning is directly exhibited in intuition rather than inferred or postulated.

Weyl declares:

> All knowledge, while it starts with intuitive description, tends toward symbolic construction.
>
> (1949, 75)[9]

This trajectory, from intuition to symbolization, defines the core dynamic of his epistemology. Yet crucially, the intuitively given is not reducible to raw sensory data. As Weyl emphasizes, its contents are "not composed of the mere stuff of perception, as many Positivists assert" (Weyl 1994, 4)[10]. Rather, they constitute the residuum of phenomenological reduction: the irreducible layer of conscious experience through which objects are genuinely presented to the subject.

From this intuitive ground, symbolic representations are constructed through *intentional acts*. The objects posited in these representations—spacetime points, fields, particles—are not things-in-themselves but intentional objects, possessing only phenomenal existence. Through reflective acts of phenomenological reduction, such transcendent constructs can be traced back to their intuitive origins. As Weyl notes in the introduction to his *Space-Time-Matter* (1952)[11], these are precisely the "intentional objects" of scientific cognition.

Weyl outlines his schema for symbolic construction as follows:

i **Starting Points**

The foundational elements of symbolic construction include:

- Basic categories of individual objects (e.g., variables $x,y,\ldots$);
- Immediately exhibited properties of these objects (e.g., $P(x)$) and primitive relations between them (e.g., $R(xy)$);
- The identity relation ("x is identical to y") among objects within the same basic category.

ii **Primitive Judgments**

Judgments that directly reflect these intuitively given properties and relations are termed simple or primitive judgments. These serve as the atomic building blocks of scientific proposition—basic statements that require no further decomposition[12].

iii **Principles of Combination**
Upon this intuitive foundation, complex judgments[13] concerning derived properties or relations are formed through iterative applications—in any order—of logical operations that Weyl calls *the principles of combination*. These include negation ("not"), identity ("is identical to"), conjunction ("and"), disjunction ("or"), existential quantification ("there is"), and variable substitution ("filling in").

These principles operate recursively on both simple and already-complex judgments, enabling the systematic construction of increasingly sophisticated scientific claims. See (Weyl 1994, Ch. 1)[14]; (Mancosu 2010, Ch. 9)[15] for a foundations of mathematics perspective on this outline.

Despite the apparent clarity of this framework, significant ambiguities remain, particularly concerning the nature and delimitation of the "basic categories," as well as the status of primitive properties and relations. What, exactly, counts as a basic category? Why must its elements be immediately given rather than conventionally stipulated?

Weyl's answer is unequivocal: these foundational elements cannot be chosen arbitrarily; they must be grasped directly through intuition. In the domain of mathematical and physical cognition, he identifies at least one indispensable basic category: that of the natural numbers. Here, individual natural numbers serve as the elementary objects, and the "immediate successor" relation $S(xy)$, expressing that y succeeds x in the number sequence, is the paradigmatic primitive relation. Both are, for Weyl, exhibited in intuition[16].

Moreover, drawing on what he calls the "a priori intuition of iteration (repeated operations) and the sequence of natural numbers,"—the capacity to grasp repeated operations and the endless progression of the number sequence—Weyl argues that the core concepts of pure number theory and even set theory can be secured without recourse to external axioms or logical machinery. As he later reflects:

> only when I had achieved certain general philosophical insights….I became firmly convinced (in agreement with Poincaré, whose philosophical position I share in so few other respects) that *the idea of iteration, i.e., of the sequence of the natural numbers, is an ultimate foundation of mathematical thought…*
>
> (Weyl 1994, 48)[17]

Each of the basic categories, in other words, carries its own constellation of immediately given structures, which together form the bedrock of scientific representation.

The Introduction of a Local Coordinate System

The Chasm Between the Intuitive and Mathematical Continua

Within Weyl's epistemological schema, the intuitively given applies to individual entities—objects that are immediately present to consciousness. Consequently, Weyl identifies "basic categories" as those that encompass individual objects, which are themselves immediately given. For example, a natural number is an individual within the basic category of natural numbers; similarly, a spacetime coincidence (such as the meeting of two worldlines) is treated as an individual within the spacetime continuum. In this framework, even a mathematical point is formally regarded as an individual member of the continuum.

Yet a crucial asymmetry arises when we compare the intuition of natural numbers with that of points in the continuum. While we can intuitively distinguish "one," "two," or "three" through acts of counting, thereby grasping natural numbers from counting discrete, individuated contents, we never directly experience a point in space or time. What we encounter instead is the intuitive continuum: temporal durations, spatial extensions, or the flowing texture of lived experience. As Weyl puts it, "exact time- or space-points are not the ultimate, underlying atomic elements" of our experience of time or space. Rather, they are understood or *constructed* based on our immediate experience of temporal durations and spatial extensions (Weyl 1994, section 6)[18].

Thus, although spacetime points function as individuals in the formal structure of physical theory, they are not given in intuition. The intuitive continuum, characterized by vagueness, imprecision, and qualitative richness, is what is genuinely accessible to consciousness. By contrast, the mathematical continuum, composed of *homogeneous, indistinguishable points*, is a product of symbolic construction. As Weyl observes:

> Certainly, the intuitive and the mathematical continuum do not coincide; a deep chasm is fixed between them.
>
> (1994, 93)[19]

This "chasm" marks the central epistemological challenge in Weyl's project: how to construct an exact, objective representation of the *world* from the inherently inexact medium of immediate experience.

The Coordinate System as the Necessary Residue

Despite the inherent vagueness and imprecision of immediate experience, Weyl insists that rational motives drive us toward a comprehension of the physical world in a "truly objective, exact, and non-qualitative" manner

(Weyl 1994, 93)[20]. To bridge the gap between the intuitive and mathematical continua, Weyl claims that the intuitive continuum "must be replaced by a four-dimensional continuum in the abstract arithmetical sense" (Weyl 1949, 113)[21]. This transition is achieved through the introduction of a coordinate system.

Crucially, Weyl does not begin with points. Instead, theoretical construction starts from intuitively graspable basic categories of extended magnitudes, for instance, time spans or spatial regions. With the aid of a continuity axiom, these can be systematically measured and ordered, eventually yielding a numerical representation. As Weyl explains,

> After choosing a definite system of coordinates…All geometric relations thus find their arithmetic-logic representation.
>
> (2009, 106)[22]

The coordinate system is thus an intentional act performed to translate intuitive experiences into precise mathematical frameworks by a rational subject driven by the imperative of objective representation.

> The coordinate system is the unavoidable residue of the eradication of the ego in that geometrical-physical world which reason sifts from the given using 'objectivity' as its standard—a final scanty token in this objective sphere that existence is only given and can only be given as the intentional contents of the process of consciousness of a pure, sense-giving ego.
>
> (Weyl 1994, 94)[23]

The introduction of a coordinate system is "unavoidable" or *necessary* because "…only reason, which thoroughly penetrates what is experientially given, is able to grasp those exact ideas" (Weyl 1994, 94)[24]. The coordinate system "eliminates" the inherent vagueness of intuitive experience with exact representations based on arithmetic and logic, a cornerstone of mathematical and physical cognition. In this sense,

> Analytical geometry as founded by Descartes is the device by which we eliminate intuitive space from constructive physics. After choosing a definite system of coordinates… All geometric relations thus find their arithmetic-logic representation.
>
> (Weyl 2009)[25]

However, a "*residue*" remains, as Weyl declares,

> [This *elimination*] does not fully succeed, and the coordinate system remains as the necessary *residue* of the ego-extinction. It is good to

> remember here that in practice two- or three-dimensional point-sets are usually given by actually putting a body or a figure drawn with pencil on paper before our eyes, and not by a logic-arithmetical construction of *set-defining* properties.
>
> (1949, 75)[26]

Importantly, the coordinate systems introduced here are not global ones. Because the intuitively given is confined to experience of localized space-time regions—what one sees, touches, or measures *here and now*—the co-ordinate system is introduced only over a local or infinitesimal region of the continuum. This aligns with the mathematical notion of the tangent space at each point, in which vectors are thought of as the vectors tangent to the continuum (or manifold) at that point. Just as differential geometry assigns linear approximations locally, Weyl's epistemology restricts symbolic construction to neighborhoods grounded in immediate intuition. The global continuum, therefore, is not intuited as a whole but patched together from local coordinatizations.

Purely Infinitesimal Geometry and the Problem of Space

In Weyl's analysis of space and geometry, only spatio-temporal coincidences and the basic geometric or spatio-temporal properties and relations within an immediate neighborhood of such coincidences possess a meaning that is "directly clear" and exhibited in intuition. As he puts it:

> …only the spatio-temporally coinciding and [basic geometrical or spatial-temporal relations within] the immediate spatial-temporal neighborhood have a directly clear meaning exhibited in intuition.
>
> (Weyl 1931)[27]

Consequently, the starting points for mathematical construction in geometry are not global structures but local relations defined in the infinitesimal vicinity of a point on a manifold. This epistemological constraint, rooted in the limits of intuitive givenness, shapes the very architecture of Weyl's geometric framework.

Unlike Riemannian geometry, Weyl's purely infinitesimal geometry only allows the comparison of lengths at the same point. The components of a vector depend on the choice of coordinate system. The determination of vector lengths relies on the choice of "calibration" or what Weyl calls a gauge. At each point of the manifold, one may freely choose this calibration, meaning that absolute lengths are not determined; only ratios of lengths between vectors at the same point possess invariant significance[28].

This dual freedom—coordinate transformations and independent rescalings of the metric—introduces a new local symmetry into the geometry: the gauge symmetry. Formally, if the squared length of a vector ξ at a point is given by

$$ds^2 = g_{\mu\nu}(x)dx^{\mu}dx^{\nu} \tag{5.1}$$

then, in addition to general coordinate transformations, the metric is subject to conformal (gauge) transformations of the form

$$g_{\mu\nu}(x) \rightarrow \bar{g}_{\mu\nu}(x) = \lambda(x)g_{\mu\nu}(x), \tag{5.2}$$

where $\lambda(x)$ is an arbitrary smooth scalar function—the gauge function. Metrics related by such a transformation are regarded as equally valid representations of the same physical-geometric situation.

The behavior of vectors under infinitesimal displacement is governed by two structures. First, the change in the components of a vector ξ_{μ} under parallel transport, a mathematical operation which leaves the vector "parallel" to itself under transport from a point P to an infinitesimally neighboring point P′ is defined by

$$d\xi^{\mu} = -\xi^{\rho}\Gamma^{\mu}_{\rho\nu}dx^{\nu}, \tag{5.3}$$

where $\Gamma^{\mu}_{\rho\nu}$ is an affine connection. A point P is said to be affinely connected to its neighborhood if there exists a coordinate system—valid in the immediate vicinity of P—in which the components of any vector remain unchanged under infinitesimal parallel displacement. Such a system is called a geodesic coordinate system at P.

Second, due to the metric's gauge dependence, the magnitude of a vector also changes under parallel transport. Specifically, the variation in length is given by

$$d\left(g_{\mu\nu}\xi^{\mu}\xi^{\nu}\right) = -g_{\mu\nu}\xi^{\mu}\xi^{\nu}\varphi_{\rho}dx^{\rho}, \tag{5.4}$$

where φ_{ρ} is a length connection or gauge field. In a manifold, a point P is considered metrically connected to its surrounding neighborhood if there exists a gauge choice (applicable to the immediate vicinity of P) that preserves the length of any vector at P when it's congruently displaced to infinitesimally nearby points. This specific gauge at point P is called a geodesic gauge (or simply a geodesic).

Combining these requirements, Weyl derives a unique connection—the Weylian connection—that respects both affine and metrical parallelism in a gauge-covariant manner:

$$\Gamma^{\mu}_{\rho\nu} = \frac{1}{2}g^{\mu\nu}\left(g_{\sigma\rho,\nu} + g_{\sigma\nu,\rho} - g_{\rho,\nu,\sigma}\right) + \frac{1}{2}\left(g^{\kappa}_{\rho}\varphi_{\nu} + g^{\kappa}_{\nu}\varphi_{\rho} - g_{\rho\nu}g^{\mu\eta}\varphi_{\eta}\right). \tag{5.5}$$

Crucially, this geometric structure is invariant not only under arbitrary coordinate transformations but also under simultaneous gauge transformations of the metric and the length connection:

$$g_{\mu\nu} \rightarrow \bar{g}_{\mu\nu} = \lambda g_{\mu\nu} \text{ and } \varphi_\rho \rightarrow \varphi_\rho = \varphi_\rho - \frac{1}{\lambda}\lambda_\rho. \tag{5.6}$$

Global geometric relations between distant regions of the manifold are then constructed by iterating these infinitesimal operations: parallel transport governed by the affine connection, together with length comparison mediated by the gauge field. In this way, Weyl builds a purely infinitesimal geometry, one that unfolds in three conceptual stages:

> from the *continuum*, which eludes closer definition, by way of *affinely connected manifolds*, to *metrical space*.
>
> (Weyl 1952, 102)[29]

This progression reflects Weyl's deeper philosophical conviction: that the continuum is not a pre-given, globally structured entity, but a constructive achievement grounded in local, intuitively accessible relations. The resulting geometry is thus not merely a mathematical formalism but an epistemological schema—one that respects the limits of intuition while enabling the rational reconstruction of objective spatial structure.

The Problem of Intersubjective Constitution

"The Mystery of the Intersubjective Communication"

The introduction of a coordinate system marks only one stage in the broader process of symbolic construction. This process, in Weyl's view, must originate in the authentic immediate experience of individual subjects. Yet such experiences are inherently private: the conscious content of subject S_1 remains directly inaccessible to subject S_2. No matter how vivid or certain S_1's intuition may be, it cannot be shared as such with another mind.

Scientific knowledge, however, cannot rest on the foundation of isolated subjectivity. It must be intersubjective—grounded not in the solipsistic immediacy of a single ego, but in a convergence of perspectives that transcends any one individual's standpoint. This raises a fundamental challenge: if symbolic construction begins with the irreducibly subjective data of immediate experience, how can genuinely objective, intersubjective knowledge emerge?

Weyl explicitly acknowledges this epistemological impasse. He writes:

> …if I furthermore accept as data the contents of the consciousness of others on equal terms with my own, thus opening myself to *the mystery of the intersubjective communication*…
>
> (Weyl 1949, 117)[30]

This "mystery" names a second chasm—one distinct from, yet parallel to, the earlier divide between the intuitive and the mathematical continuum. While the first chasm concerns the transition from vague, qualitative experience to exact symbolic representation, the second chasm lies between the givenness of "I" and the accessibility of "others." To achieve what Weyl calls "the 'truly objective', exact, non-qualitative" world, both chasms must be bridged. And it is our "rational motives"—the drive toward objectivity—that compel us to navigate this dual terrain.

"The Perspective View," "the Photogrammetry View," and "Eliminating the Body"

Weyl vaguely sketches three possible strategies for achieving intersubjective constitution. The first two he associates with familiar perceptual paradigms: the "perspective view" and the "photogrammetry view." The third, more radical approach, involves "eliminating the body."

> ...Instead of constructing the perspective view which a given body offers from a given point of observation, or conversely constructing the body from several perspective images, as it is done in photogrammetry, we might eliminate the body...
>
> (Weyl 1949, 117)[31]

In the perspective view, a single subject S_1 constructs a representation of the world based exclusively on her own immediate experiences. This view is anchored in the spatio-temporal encounters between S_1's body and the environment—what Weyl elsewhere calls "spatio-temporal coincidences." The resulting model is inherently egocentric: it reflects the world as it appears from S_1's unique vantage point P_1, with no built-in access to the perspectives of other subjects S_i and $i \neq 1$[32].

The photogrammetry view, by contrast, seeks to overcome egocentrism by synthesizing multiple perspective views. Just as photogrammetry reconstructs a three-dimensional object from a set of two-dimensional images taken from different angles, this approach attempts to construct a transcendent object—a shared "body" in space—by integrating the diverse observational records of multiple subjects. The underlying assumption is that these disparate perspectives all refer to a common external reality, which serves as the ontological anchor for intersubjectivity.

Weyl does not explicitly state why he finds both approaches inadequate, but plausible criticisms can be inferred. The perspective view risks collapsing into solipsism: if knowledge is confined to one subject's bodily-mediated experiences, and no principled connection exists between different subjects' streams of consciousness, then the very possibility of a shared world becomes philosophically precarious. There is no bridge from S_1's world to S_2's—only a collection of disconnected monads.

The photogrammetry view, while seemingly more promising, smuggles in an unexamined metaphysical commitment: it presupposes the existence of a pre-given, observer-independent object that all subjects are observing. But for Weyl who was deeply influenced by transcendental idealism, such an assumption is illegitimate. In his schema, objects are not given prior to constitution; they are constituted through the interplay of intuition, symbolism, and rational norms. To posit a common body as the foundation of intersubjectivity would thus invert the proper order of epistemic priority.

In response, Weyl proposes a third path: "eliminating the body." He reformulates the problem as follows:

> ...we might eliminate the body and formulate the problem directly as follows: let A, B, C each represent a consciousness bound to a point body, and let K be a solid contained in their field of vision. The task is to describe the lawful geometrical connections between the three images which each one of the three persons A, B, C receives of K and of the locations of the other two persons. This procedure would be more unwieldy; in fact it be bound to fail on account of the limitations and gaps in any single consciousness as compared to the complete real world.
>
> (1949, 117)[33]

This approach dispenses with the notion of a substantial, extended body and treats observers as idealized point-consciousnesses. The goal is no longer to reconstruct an external object, but to articulate the invariant geometric relations among the perceptual fields of different subjects.

Yet this solution appears to sidestep the very problem it aims to resolve. By reducing subjects to interchangeable points labeled A,B,C, Weyl abstracts away from the qualitative or quantitative interiority that makes intersubjectivity mysterious in the first place. The "mystery of intersubjective communication" lies precisely in the fact that other minds are not mere loci of perception but centers of lived experience—each with its own horizon of meaning, intentionality, and temporal flow. To "eliminate the body," then, is to erase the very concrete conditions that give rise to problems in intersubjective communication and recognition, and mutual understanding. Therefore, the approach may yield formal symmetries, but it fails to account for the phenomenological depth of intersubjective life. Weyl himself seems aware of this limitation. He concedes that such a procedure "would be more unwieldy; in fact it would be bound to fail on account of the limitations and gaps in any single consciousness as compared to the complete real world" (1949, 117).

While Weyl doesn't offer a definitive solution to "the mystery of intersubjective communication," his schema might hold a key to bridging the gap between subjective and objective experience. To explore this possibility, I delve into two key pairs of dualities within the Weylian framework, with the aid of conceptual tools which I introduced for my own discussions on publicity.

Two Pairs of Dualities

The Dual Nature of the Intuitively Given and of the Ego

The foundation of all theoretical construction, according to Weyl, lies in what is intuitively given, yet this starting point itself possesses a dual nature. On one hand, it is *given*: it arises through the concrete encounters between a subject (an ego) and the world. Spatial extents, temporal durations, and other primitive individuals are presented to the ego via its *embodied interaction* with reality. On the other hand, it is *intuitive*: the contents of this givenness—the basic categories, their intrinsic properties, and the relations among them—are disclosed not through inference but through intuition, which yields immediate, certain judgments.

Thus, the intuitively given is not a passive datum but a dynamic unity of the sensible (what is sensed) and the cognitive (what is understood)[34]. As Weyl puts it:

> One has to acknowledge according to Kant and Fichte: I am originally endowed with the faculty of intuition as well as sensation. A thing can exist for me only in the indissoluble unity of sensation and intuition…
>
> (2009, 102)[35]

This duality is not accidental but essential. For Weyl, the intuitively given cannot be split into raw sensation on one side and conceptual overlay on the other; rather, meaning is gradually generated in the very act of intuitive apprehension.

Both faculties—sensibility and cognition—belong to the ego, or subject. The ego's empirical body serves as the interface between the physical world and inner experience: it registers sensory data, locates the subject in space and time, and grounds perception in finitude. In this capacity, the ego is an empirical entity, born of a mother and destined to die (Weyl 2009, 197 and 215)[36], coexisting with countless other human egos within the natural order.

Yet the ego is also more than a biological organism. It functions as a transcendental agent—the source of meaning-giving acts that interpret the intuitively given, formulate primitive and complex judgments, and construct symbolic representations that reach beyond immediate experience. In this role, the ego does not merely receive the world; it constitutes it. As Husserl writes:

> I myself, as transcendental ego, "constitute" the world, and at the same time, as soul, I am a human ego in the world. The understanding which prescribes its law to the world is my transcendental understanding, and it forms me, too, according to these laws…
>
> (1970, 202)[37]

Accordingly, the ego must be understood as both empirical and transcendental: situated within the causal nexus of nature, yet simultaneously the origin of normative structures that shape scientific knowledge. This dual status mirrors the double nature of the intuitively given itself, anchored in bodily encounter, yet illuminated by cognitive insight.

Crucially, the transcendental ego accesses the meaning of what is given to the empirical ego through intuition, although this access remains subjective: what is intuitively given to subject S_1 is uniquely hers and cannot be directly shared with others S_i, here $i \neq 1$. The empirical ego perceives *the given*; the transcendental ego *intuitively* interprets. Together, they form the two poles of a single, bifurcated subjectivity—one immersed in the world, the other constituting it.

The Sphere of Being and the Sphere of Meaning

Symbolic representations arise through iterative operations governed by what Weyl calls "principles of judgment combination." These principles do not operate in a vacuum; they are grounded in the intuitively given, which serves as the indispensable foundation for all theoretical construction. The intuitively given includes the basic category of individuals, along with their primitive properties and relations, and functions as the elementary building block from which more complex structures are erected.

For each subject S_i, this foundational layer constitutes what I term a composite "sphere of being." At its most fundamental level, this sphere comprises the natural numbers together with the immediately exhibited successor relation *S*(*xy*)—a structure Weyl analyzes in detail in *The Continuum* (1994, 24–28)[38]. Within this sphere, any relational configuration among objects can be systematically positioned relative to the intuitively given through formal construction. The term "composite" is crucial: this sphere integrates two interwoven dimensions what I term: *a sphere of constituted things*, derived or constructed by subjects S_i from the basic category, and *a sphere of meanings*, generated through the interpretation of primitive and derived properties and relations by the sense-giving ego or transcendental ego.

Weyl seems to draw a sharp ontological distinction between these two realms. He writes:

> The world exists only as that met with by an ego, as one appearing to a consciousness; the consciousness in this function does not belong to the world, but stands out against the beings as the sphere of vision, of meaning, of image, or however else one may call it …
>
> (2009, 83)[39]

To model this structure geometrically, Weyl proposes an analogy:

> The objects, the subjects, and the way an object appears to a subject, I model by the point, the coordinate system, and the coordinates of a point with reference to a coordinate system in geometry…Here objects (points) and subjects (coordinate system=triplets of points) belong to the same *sphere of reality*. The appearances of an object, however, lie in another sphere, in the *realm of number*.
>
> (2009, 216)[40]

Following the establishment of a *local* coordinate system, a numerical framework, *numerical meaning or sphere of meaning*, is constructed. Through repeated applications, or *iterations*, of the principles of combination, a realm of *constructed objects*, *sets* (representing properties) and *functional connections* (representing relations)[41], *a sphere of being with derived meaning*, emerges.

Yet a crucial epistemic asymmetry remains. While intuitive experience delivers knowledge with immediate certainty, symbolic constructs, though rooted in intuition, do not inherit this epistemic security. Intuitive acts are temporally finite; they occur within the fleeting present of lived experience. Symbolic operations, by contrast, achieve durability by being transferred onto representing symbols that are "time-resisting and simultaneously serve the purpose of preservation and communication" (Weyl 2009, 119)[42]. In this transfer, symbolic constructs gradually detach from their intuitive origins, acquiring a degree of autonomy.

Consequently, the resulting sphere of constructed objects, properties, and relations and the sphere of meaning is constituted upon the intuitively given, even though it might not be a faithful mirror of empirical reality. Rather, it constitutes a realm of *derived possibilities constructed by iterations of mathematical principles*—more precisely, "an ordered manifold of possibilities producible according to a fixed procedure and open into infinity" (Weyl 2009, 69, 118, 119, and 185)[43].

In light of the conceptual architecture developed in earlier chapters, the sphere of being and the sphere of meaning can now be articulated as two interrelated yet ontologically distinct realms within a dual-structured ontology.

On one hand, there is the sphere of objects, progressively constructed through the iterative application of mathematical principles to the elements of the basic category. On the other, there is the sphere of meanings, constituted by the continuous enrichment of numerical interpretations assigned to local objects, especially those defined within infinitesimal regions, through the repeated application of these same mathematical principles to elements of the basic category.

Crucially, the basic category of individual objects, such as spatial extents and temporal durations in infinitesimal neighborhoods, or other primitive entities, forms the domain of beings that genuinely interact with the empirical ego "I." These are given directly in intuition; their presence is immediate, finite, and epistemically certain. In contrast, the constructed objects that emerge from iterative symbolic operations—though rooted in this intuitive substrate—never enter into direct experiential contact with the ego. They are not given but posited: ideal, possible entities brought forth by the transcendental activity of the "I."

Correspondingly, the meanings immediately conferred in the ego's direct encounter with an object carry immediate certainty; they are inseparable from the intuitive act itself. Yet as mathematical operations accumulate—layering complex judgments upon primitive meanings—the resulting constituted meanings no longer refer to actual beings. Instead, they articulate possible objects within an ever-expanding manifold of formal possibilities. These meanings belong not to the sphere of being, but to the sphere of meaning: a structured realm of ideality, open-ended and infinite in scope, yet always traceable back to its intuitive origin.

Thus, while the ego remains firmly anchored in the concrete world through its empirical embodiment, its transcendental function simultaneously projects it into a symbolic realm—one not of actuality, but of constructible possibility.

Constitution of Intersubjectivity in Spacetime Dimension

The desire to comprehend the world in a "truly objective, exact, non-qualitative" manner—what Weyl calls "rational motives"—leads naturally to the adoption of coordinate systems. Yet such systems primarily satisfy only the exact and non-qualitative dimensions of this aspiration. The deeper demand for "true objectivity," that is, intersubjective validity, remains unfulfilled by the introduction of coordinate systems alone.

Introducing a local coordinate system does not automatically yield intersubjective understanding. The fundamental obstacle lies in the inherent subjectivity of the epistemic starting point: all knowledge originates in the intuitive horizon of a concrete ego. This first-person givenness stands in tension with the very rational motives that call for a shared, trans-subjective account of reality.

Scientific representation, unlike purely private experience, aims precisely at intersubjectivity. It seeks to articulate the world not as an "ordered manifold of possibilities" confined to a single subject "I," or S_1, but as a shared manifold that integrates the standpoints of a plurality of egos $\{S_i\}$, here $i \neq 1$.

This intersubjective manifold presupposes a crucial condition: that S_1 can access and meaningfully interpret the sphere of meaning given by another subject S_i ($i \neq 1$), and *vice versa*. Mutual comprehension requires that each ego be able to grasp the meanings and intuitions that constitute the other's world.

When such reciprocity is achieved, the intuitive experiences of one subject become functionally equivalent to those of others within the intersubjective order. Drawing on Weyl's geometric analogy, this equivalence manifests as *the homogeneity of spatial structure in the infinitesimal neighborhoods where different bodily ego resides*. Just as the geometric relations around any point in a homogeneous space are indistinguishable from those around any other, so too must the constitutive structures of local experience be uniform across subjects if genuine intersubjectivity is to arise.

As Weyl writes:

> …transition to homogeneous space, where the bodily ego takes on a position on equal terms with other bodies, this is accomplished by the possibility of walking toward the distant horizon of centered space, by the free mobility of our own body in space and by the intentions of our will directed toward such motion. Not before this last step do I become capable of imagining myself as being in the position of another person, only this space may he thought of as the same for different subjects, it is a medium necessary for constructing an intersubjective world.
>
> (2009, 101)[44]

In light of this, one may say: each concrete subject occupies a point, or more precisely, an infinitesimal region in physical spacetime, and this occupation evolves along a worldline as time progresses. Another concrete subject similarly occupies its own spacetime region, tracing its own trajectory through the manifold.

Weyl's requirement of a homogeneous space, where "the bodily ego takes on a position on equal terms with other bodies," thus entails not only geometric symmetry but also phenomenological parity among egos. It demands that the meanings constituted by different subjects through their respective local reference frames and embodied perspectives be commensurable within a common structure.

Consequently, two kinds of transformations must leave aspired intersubjective judgments invariant: changes in convention, such as a rescaling of the unit of length (a change in the scalar factor defining measurement); and changes in reference frame, such as shifts in spatial orientation or state of motion.

Neither alteration should affect judgments concerning the intersubjective or objective properties or relations—local or global—of the spacetime manifold. These invariances correspond precisely to the transformations between subjects, whether conceived as physical variations (e.g., moving to a new location) or mental/conventional shifts (e.g., adopting a different ruler or coordinate chart).

Thus, the homogeneity of space functions as the formal pre-condition under which the "I-judgment" of one ego can be constructed as the "we/it-judgment"

of a shared scientific world, enabling the symbolic construction of an intersubjective reality grounded in, yet transcending, individual intuition.

I believe the following passage from Weyl strongly resonates with the interpretation developed above:

> ...the objective state of affairs contains all that is necessary to account for the subjective appearances. There is no difference in our experiences to which there does not correspond a difference in the underlying objective situation (a difference, moreover, which is invariant under arbitrary coordinate transformations). It comprises as a matter of course the body of the ego as a physical object.
>
> (1949, 116)[45]

This statement can be read not merely as a metaphysical assertion about intersubjective or objectivity, but as a phenomenological principle of completeness: every variation in subjective experience must be grounded in an invariant feature of the objective world. In other words, the realm of intersubjectivity or objectivity must be rich enough and structured precisely enough to encode all possible subjective perspectives as legitimate manifestations of a single underlying reality.

From this perspective, the initial intuitive experiences of the ego, though immediate and certain, are not endpoints, but starting points for a process of enrichment. Driven by rational motives and the demand for intersubjectivity, the ego begins to consider not only its own local intuitions but also to anticipate the potential intuitive horizons of other subjects. This reflective expansion transforms the originally private "I-judgment" into a node within a broader intersubjective manifold of possibilities—possible entities, possible relations, and possible properties.

Consequently, within this organized structure of possibilities, two crucial shifts occur: distant properties and relations are no longer taken as given but are symbolically constructed through mathematical principles applied to local data; and even the local region itself—once anchored exclusively in the unique intuition of S_1—is reconceived as a multiplicity of *potential* intuitive standpoints, each corresponding to *a possible embodied observer* situated at that spacetime location.

In this light, introducing an additional conformal structure in the purely infinitesimal geometry, beyond the affine connection of Riemannian geometry, can be understood as a deliberate methodological move to absorb at least one source of conventional indeterminacy—the possible conventionality in calibration—into the very fabric of geometric representation. The length of a vector, for instance, is no longer absolutely fixed at a point; instead, it is defined only up to a local scale factor, reflecting the fact that measurement conventions (e.g., choice of unit rod) are not dictated by nature but chosen by observers.

Thus, the invariance requirements for genuine objectivity become twofold: *general covariance* (invariance under arbitrary coordinate transformations as already required by general theory of relativity), ensuring that the description does not depend on the observer's spatial or temporal labeling scheme; and *gauge (or conformal) invariance*, ensuring that it does not depend on the observer's choice of local scale or calibration standard.

Only when both conditions are satisfied can we speak of judgments that are truly intersubjective—not tied to any particular ego's frame of reference or measuring convention, but valid for any embodied subject capable of traversing spacetime and performing measurements. These invariant structures constitute the symbolic medium through which the private "I" enters into the public "we/it."

In light of the symbolic formulation introduced in Chapter 4 on the construction of conjectural "we/it," consider an "I-form" representation of spacetime or a physical law thereof, articulated by a specific ego "I":

$$f^{\text{``}I\text{''}}\left(S^{\text{``}I\text{''}}, O, \{C_1, C_2, \ldots, C_m, \ldots\}, E\right),$$

where O now denotes the object domain (e.g., spacetime objects).

Weyl's spacetime geometry posits that the conventionality inherent in an observer's choice of coordinate systems and measurement calibrations reflects variability in the subject's methodological or conceptual framework—not in the physical world itself. To fulfill the "rational motives" of achieving a truly objective representation and overcoming the "mystery of intersubjective communication," this "I-form" judgment must first be reformulated into a "we/it" form:

$$f^{\text{``}we/it\text{''}}\left(S^{\text{``}we/it\text{''}}, O^{\text{``}we/it\text{''}}, \{C_1, C_2, \ldots, C_m, \ldots\}^{\text{``}we/it\text{''}}, E^{\text{``}we/it\text{''}}\right),$$

with the crucial requirement that truth be invariant under variations in coordinate systems and rescalings of rulers or clocks. That is, for any two instances of the possible subjects $S_{\mu}^{\text{``}we/it\text{''}}$ and $S_{\nu}^{\text{``}we/it\text{''}}$—who may adopt different reference frames or calibration conventions—the following holds:

$$\left(S_{\mu}^{\text{``}we/it\text{''}}, O^{\text{``}we/it\text{''}}, \{C_i\}^{\text{``}we/it\text{''}}, E^{\text{``}we/it\text{''}}\right) \rightarrow \left(S_{\nu}^{\text{``}we/it\text{''}}, O^{\text{``}we/it\text{''}}, \{C_i\}^{\text{``}we/it\text{''}}, E^{\text{``}we/it\text{''}}\right),$$

Here, only the subject term carries subscripts carry subscripts (μ, ν), since the variability in question stems solely from differences in observers' choices.

This invariance condition directly corresponds to the mathematical structure of the Weylian connection, which, as a covariant derivative, is required to be invariant not only under arbitrary coordinate transformations (diffeomorphisms) but also under local gauge (scale) transformations. In this sense, the additional structure and invariance introduced in Weyl's spacetime theory may be understood as aspiring to ensure that the form of physical law remains identical across all possible subjects employing different conventions for reference frames and measurement scales.

Nevertheless, the above invariance requirement comes at a cost. The conformal structure—by relinquishing absolute local metrical determinacy—introduces a kind of indeterminacy into the geometric foundation. One might say that this is the price Weyl pays for intersubjectivity, and, more broadly, for publicity: the condition that scientific knowledge be accessible, verifiable, and communicable across a plurality of egos. Objectivity, in this view, is not the elimination of subjectivity, but its systematic integration into a shared symbolic order.

It is true that Weyl's theory failed as a description of physical reality. Yet if we adopt the interpretation advanced in Chapter 4—according to which public or intersubjective representation is not a mirror of reality-in-itself, but rather the construction of conjectural "we/it-judgments" grounded in shared operational practices—then this so-called failure does not reflect a flaw in Weyl's underlying strategy. Rather, it stems from a misplaced expectation: namely, the assumption that conventions governing measurement—such as the choice of unit length or "ruler"—could physically alter the intrinsic properties of spacetime, whether locally or globally.

From this perspective, the empirical falsification of Weyl's geometric model reveals not a defect in his epistemological vision, but a limitation arising from ontological overreach. The error lay not in his attempt to ground intersubjectivity in geometric invariance—an ambition deeply aligned with the "rational motives" he identified—but in conflating a feature of our representational practices with a feature of the world itself. Specifically, Weyl treated the freedom of scale, which properly belongs to the domain of convention and intersubjective coordination among observers, as though it corresponded to a real physical field capable of influencing material processes. Yet physical reality appears to tell a different story: while conventions undoubtedly govern our choice of rulers and clocks for measuring length and time, nature itself seems to furnish intrinsic standards—absolute scales rooted in fundamental physical constants (such as Planck constant) and processes (such as atomic transitions). These underlie the very possibility of stable measurement and thus constrain, rather than license, arbitrary gauge freedom in the physical realm.

Once we decouple the epistemic function of the conformal structure—its role in securing publicity, translatability, and mutual intelligibility among embodied observers—from the ontic claim that scale is physically variable, Weyl's framework regains its philosophical force. Freed from the burden of direct physical instantiation, the conformal structure emerges not as a failed hypothesis about nature, but as a formal embodiment of scientific openness: it encodes the requirement that objective knowledge remain invariant under legitimate variations in perspective, instrumentation, and convention.

Thus, the enduring value of Weyl's infinitesimal geometry lies in its demonstration of how the demand for intersubjectivity can actively shape the architecture of symbolic representation. Far from being a mere technical device, the conformal structure becomes a structural expression of plurality—acknowledging that each observer inhabits their own intuitive horizon, yet affirming that these diverse standpoints can converge on a common, publicly accountable world through invariant relational forms.

In this light, Weyl's "failure" is better understood as a productive tension: one that sharply illuminates the boundary between the conditions necessary for public representation and the independent character of physical reality. It is precisely this tension—between what must be shared to count as objective and what simply is—that makes his infinitesimal geometry an exemplary case study for the framework developed in this work.

Notes

1 Also, Norman Sieroka provides an interpretation of Weyl's understanding of the problem of intersubjectivity on the basis of a Medicus-Fichtean theory of intersubjectivity as a kind of precondition for all (individual) subjectivity (See Sieroka, Norman. "Neighbourhoods and intersubjectivity: analogies between Weyl's analyses of the continuum and transcendental-phenomenological theories of subjectivity." In *Weyl and the Problem of Space*. Edited by Julien Bernard and Carlos Lobo, 99–122. Switzerland: Springer, 2019. While I acknowledge this line of interpretation, my aim here is not primarily exegetical. Rather, I use Weyl's theoretical framework as a lens through which to examine the problem of publicity (or intersubjectivity) in the construction of in spacetime dimensions.

2 O'Raifeartaigh, Lochlainn and Norbert Straumann. "Gauge theory: History origins and some modern developments." *Reviews of Modern Physics*, 72 (January 2000): 1–23; Goenner, Hubert. "On the history of unified field theories." *Living Reviews in Relativity*, 7 (December, 2004); Ryckman, Thomas, "Hermann Weyl, the Gauge principle, and symbolic construction from the 'Purely Infinitesimal,'" In *One Hundred Years of Gauge Theory*. Edited by Silvia De Bianchi and Claus Kiefer, 161–183. Switzerland: Springer, 2020.

3 Terms and motifs from Husserl's phenomenology appear frequently in Weyl's writings, and scholars have explored the affinities between Weyl's "purely infinitesimal" approach and transcendental-phenomenological themes, see Feist, Richard. "Husserl and Weyl: phenomenology, mathematics, and physics." In *Husserl and the Sciences*. Edited by Richard Feist, 153–172. Ottawa: University of Ottawa Press, 2004; and Ryckman, Thomas. *The Reign of Relativity* (especially chapters 5 and 6). New York: Oxford University Press, 2005; etc. Nevertheless, Weyl was no uncritical follower of Husserl. As he later remarked, "a number of Husserl's theses become demonstrably false when translated into the context of this analogy—something which, it appears to me, gives serious cause for suspecting them" (Hermann, Weyl. "Insight and reflection." In *Mind and Nature: Selected Writings on Philosophy, Mathematics, and Physics*. Edited by Peter Pesic, 204–221. Princeton, NJ and Oxford: Princeton University Press, 2009. Also see Bell (2004)'s brief discussion on Weyl's divergence from Husserl. Bell, John. "Hermann Weyl's later philosophical views: his divergence from Husserl." In *Husserl and the Sciences*. Edited by Richard Feist, 173–185. Ottawa: University of Ottawa Press, 2004. Indeed, Weyl drew eclectically from multiple philosophical traditions, including

Fichte, Kant, and even Leibniz, crafting a distinctive epistemological stance that resists simple categorization.

4 Weyl, Hermann. *The Continuum*. New York: Dover Publications, 1994.

5 Tieszen, Richard. "The philosophical background of Weyl's mathematical constructivism." *Philosophia Mathematica*, 8 (2000): 274–301; Ryckman, Thomas. *The Reign of Relativity*. New York: Oxford University Press, 2005.

6 Weyl, Hermann. *The Continuum*. New York: Dover Publications, 1994.

7 Weyl, Hermann. *Philosophy of Mathematics and Natural Science*. Princeton, NJ: Princeton University Press, 1949.

8 Weyl, Hermann. *The Continuum*. New York: Dover Publications, 1994. By "more or less arbitrarily axiomatized systems," Weyl is primarily critiquing a certain view in the foundations of mathematics that treats mathematics as nothing more than a meaningless game of symbols, in which definitions and axioms for constructing increasingly complex mathematical objects are chosen purely by convention.

9 Weyl, Hermann. *Philosophy of Mathematics and Natural Science*. Princeton, NJ: Princeton University Press, 1949.

10 Weyl, Hermann. *The Continuum*. New York: Dover Publications, 1994.

11 Weyl, Hermann. *Space-Time-Matter*. New York: Dover Publications, 1952.

12 Primitive judgments here may correspond to the first-person "I-form" direct judgment discussed in previous chapters.

13 Complex judgments here correspond to "judgments derived from immediate experience" in previous chapters.

14 Weyl, Hermann. *The Continuum*. New York: Dover Publications, 1994.

15 Mancosu, Paolo. *The Adventure of Reason: Interplay Between Philosophy of Mathematics and Mathematical Logic, 1900-1940*. New York: Oxford University Press, 2010.

16 For Weyl's discussions on natural number, see section 5 of Weyl (1994). Weyl, Hermann. *The Continuum*. New York: Dover Publications, 1994.

17 Weyl, Hermann. *The Continuum*. New York: Dover Publications, 1994.

18 Weyl, Hermann. *The Continuum*. New York: Dover Publications, 1994.

19 Weyl, Hermann. *The Continuum*. New York: Dover Publications, 1994.

20 Weyl, Hermann. *The Continuum*. New York: Dover Publications, 1994. Here underlies a divergence between my understanding of objectivity and Weyl. In previous, I restrain myself from talking about construct objective external world, although I argue that the conception of an objective world would still require publicity as a necessary condition.

21 Weyl, Hermann. *Philosophy of Mathematics and Natural Science*. Princeton, NJ: Princeton University Press, 1949.

22 Weyl, Hermann. "Mind and nature." In *Mind and Nature: Selected Writings on Philosophy, Mathematics, and Physics*. Edited by Peter Pesic, 83–161. Princeton, NJ and Oxford: Princeton University Press, 2009.

23 Weyl, Hermann. *The Continuum*. New York: Dover Publications, 1994.

24 Weyl, Hermann. *The Continuum*. New York: Dover Publications, 1994.

25 Weyl, Hermann. "Mind and nature." In *Mind and Nature: Selected Writings on Philosophy, Mathematics, and Physics*. Edited by Peter Pesic, 83–161. Princeton, NJ and Oxford: Princeton University Press, 2009.

26 Weyl, Hermann. *Philosophy of Mathematics and Natural Science*. Princeton, NJ: Princeton University Press, 1949.

27 Weyl, Hermann. "Geometrie und Physik." *Die Naturwissenschaften*, 19 (1931): 49–58. Ryckman, Thomas. *The Reign of Relativity*. New York: Oxford University Press, 2005.

28 For Weyls' discussions on his purely infinitesimal geometry, see Chapter II of Weyl (Weyl, Hermann. *Space-Time-Matter*. New York: Dover Publications, 1952). For other discussions on Weyl's purely infinitesimal geometry and Weylian theory of spacetime, see Pauli (Wolfgang, Pauli. *Theory of Relativity*. New York: Dover Publications, 1981), Bell and Korté (John, Bell and Herbert Korté. "Hermann Weyl."

In *Stanford Encyclopedia of Philosophy*, (Winter 2020 Edition). Edited by Edward N. Zalta. 2015), etc. Goenner [Goenner, Hubert. "On the history of unified field theories." *Living Relativity*, 2 (2004)] is also a nice introduction with discussions on various reactions to Weyl's theory.

29 Weyl, Hermann. *Space-Time-Matter*. New York: Dover Publications, 1952.

30 Weyl, Hermann. *Philosophy of Mathematics and Natural Science.* Princeton, NJ: Princeton University Press, 1949.

31 Weyl, Hermann. *Philosophy of Mathematics and Natural Science.* Princeton, NJ: Princeton University Press, 1949.

32 "The perspective view" corresponds to the first-person "I-form" direct judgment in last chapter.

33 Weyl, Hermann. *Philosophy of Mathematics and Natural Science.* Princeton, NJ: Princeton University Press, 1949.

34 Folina (2008) thinks that, like Husserl, Weyl takes intuition as both the faculty of sensibility and the faculty of understanding (Folina, Janet. "Intuition between the analytic-continental divide: Hermann Weyl's philosophy of the continuum." *Philosophia Mathematica*, 16 (2008): 25–55). Anyway, for Weyl, the intuitively given has an indissoluble double aspect, see the quote below.

35 Weyl, Hermann. "Mind and nature." *In Mind and Nature: Selected Writings on Philosophy, Mathematics, and Physics*. Edited by Peter Pesic, 83–161. Princeton, NJ and Oxford: Princeton University Press, 2009.

36 Weyl, Hermann. "Insight and reflection." In *Mind and Nature: Selected Writings on Philosophy, Mathematics, and Physics*. Edited by Peter Pesic, 204–221. Princeton, NJ and Oxford: Princeton University Press, 2009.

37 Husserl, Edmund. *The Crisis of European Sciences and Transcendental Phenomenology*. Evaston: Northwest University Press, 1970.

38 Weyl, Hermann. *The Continuum*. New York: Dover Publications, 1994.

39 Weyl, Hermann. "Mind and nature." *In Mind and Nature: Selected Writings on Philosophy, Mathematics, and Physics*. Edited by Peter Pesic, 83–161. Princeton, NJ and Oxford: Princeton University Press, 2009.

40 Weyl, Hermann. "Insight and reflection." In *Mind and Nature: Selected Writings on Philosophy, Mathematics, and Physics*. Edited by Peter Pesic, 204–221. Princeton, NJ and Oxford: Princeton University Press, 2009.

41 Each property corresponds to a set, and each relation corresponds to a functional connection. See section 4 of Weyl, Hermann. *The Continuum*. New York: Dover Publications, 1994.

42 Weyl, Hermann. "Mind and nature." *In Mind and Nature: Selected Writings on Philosophy, Mathematics, and Physics*. Edited by Peter Pesic, 83–161. Princeton, NJ and Oxford: Princeton University Press, 2009.

43 Weyl, Hermann. "The open world", "Mind and nature," and "man and foundation of science." In *Mind and Nature: Selected Writings on Philosophy, Mathematics, and Physics*. Edited by Peter Pesic, 34–82, 83–161, and 175–193. Princeton, NJ and Oxford: Princeton University Press, 2009.

44 Weyl, Hermann. "Mind and nature." In *Mind and Nature: Selected Writings on Philosophy, Mathematics, and Physics*. Edited by Peter Pesic, 83–161. Princeton, NJ and Oxford: Princeton University Press, 2009.

45 Weyl, Hermann. *Philosophy of Mathematics and Natural Science*. Princeton, NJ: Princeton University Press, 1949.

Conclusion

Scientific objectivity is often upheld and sometimes even idealized as a central norm of scientific cognition. Yet it faces persistent challenges, not only from the plurality of its conceptions, which frequently stand in tension with one another, but also from the historical contingency of its emergence and evolution. My aim in this book is neither to propose a single definition that unifies all existing notions of scientific objectivity nor to defend any particular traditional or novel account as uniquely correct. Rather, I seek to establish a deceptively simple yet profoundly consequential claim: publicity is a necessary condition for scientific objectivity.

By "publicity," I mean the feature of a judgment that renders it accessible, open to critical assessment, and capable of being endorsed by multiple epistemic agents, not as passive recipients of information but as active participants in a shared epistemic space. Across the spectrum of scientific objectivity, whether conceived through intersubjectivity, procedural or mechanical reliability, perspective independence, or fidelity to the phenomena themselves, the requirement of publicity consistently reappears as a foundational prerequisite. Without it, even the most rigorously derived result remains confined to a private horizon and thus fails to qualify as scientifically objective.

In this sense, my approach is transcendental, not in the Kantian sense of uncovering the conditions of possible experience but in identifying publicity as a necessary condition for any judgment to count as scientifically objective. This does not presuppose consensus on how objectivity is achieved; rival theories may offer conflicting accounts of what makes a claim objective. What they share, however, is a minimal but indispensable commitment: that objective knowledge must, in principle, be open to scrutiny and validation by more than one epistemic agent; there is *an agreement that people need at least some agreements*. Publicity, then, does not prescribe a single model of objectivity; rather, it establishes the common ground without which no such path could even begin. It is this minimal norm of multi-subjective accessibility that I foreground as the transcendental condition of scientific objectivity.

Importantly, publicity does not guarantee truth or full objectivity. A claim may enjoy wide acceptance and yet be profoundly mistaken, reflecting a

DOI: 10.4324/9781003783787-7

collective illusion, a culturally entrenched bias, or a dominant paradigm temporarily insulated from critique. Publicity alone cannot ensure correspondence with reality, nor can it eliminate the influence of human interests, cognitive limitations, or social structures. But it is nonetheless necessary, because objective knowledge cannot emerge from solitary reflection in total isolation. Scientific inquiry demands a community of inquirers, not because consensus validates truth, but because only through sustained mutual engagement can we move beyond the confines of individual perspective.

This insight finds support in twentieth-century philosophy of language. As Ludwig Wittgenstein claimed, meaning is inherently public; linguistic expressions gain their significance through shared uses. Similarly, Donald Davidson argued that objectivity arises not in the privacy of inner thought but through interpersonal interpretation—the triangulation among speaker, interpreter, and world. In this light, publicity is not merely a practical convenience for science; it is constitutive of both meaning and objectivity. Without it, scientific claims lose not only their warrant but their very intelligibility.

This leads to a deeper realization: the value of publicity or objectivity lies not only in the accessibility of knowledge but also in its orientation toward inclusivity, accountability, and mutual understanding. We trust science not because it speaks from a mythical "view from nowhere," but because it strives to speak from *somewhere* in a way that *invites others into the conversation*. At its best, scientific objectivity is not the erasure of subjectivity but its responsible transformation into judgments that are open, contestable, and structured so that others, regardless of background or standpoint, can engage with them critically and build upon them collaboratively. Objectivity, then, is less a final state to be attained than an ongoing practice to be cultivated: a shared commitment to reasoning together across difference.

Yet this transformation is neither automatic. To bridge the gap between one's own "I" and the many possible "I"s of others, the original observer must do more than assert findings. She must step back from her egoistic standpoint and deliberately construct a space of "possible subjects"—not as projections of her own imagination but as disciplined anticipations of legitimate differences in perception, convention, instrumentation, and interpretation. This act is not an expansion of the egocentric self; it is a relinquishment of epistemic sovereignty. It is the conscious effort to put oneself in others' shoes and to say: "This is what I see—but here is how it might appear to you, and here is why we might still agree."

This path from private insight to public knowledge is not a straight line but a structured projection, forged through two complementary acts of imagination. Outwardly, the scientist asks: "How might others differ from me in their background, tools, or expectations?" Inwardly, she reflects: "If others were in my place, how might their judgments diverge; and what would I need to show them to make my claim intelligible?" Only through this dual movement can a conjectural "we/it" judgment take shape—one that is not imposed from above, but offered as an invitation; not assumed as given, but explicitly framed for testing. And only when such a judgment survives encounter with actual others through replication, critique, reinterpretation, or collaborative refinement does it earn the status of public knowledge. Publicity, then, is not conferred by method alone; it is ratified through reception.

Pursuing this communicative openness inevitably introduces what may appear to be redundant: surplus structure, additional assumptions, explanatory scaffolding, or clarifications that a solitary thinker would never require. To the lone observer, such elaboration seems superfluous or even clutter. Yet for anyone seeking to communicate across differences in conceptual frameworks or physical situatedness, this very "excess" becomes a strategic necessity. Just as fail-safes are built into critical engineering systems, not because engineers expect total failure, but because they recognize that unforeseen stressors are inevitable, so too does the apparent redundancy in scientific communication function as an epistemic safeguard. These "extra" elements in representations are not marks of inefficiency; they are deliberate investments in robustness, designed to preserve intelligibility and enable uptake amid the unpredictability of human variation, linguistic ambiguity, and environmental contingency.

Crucially, the deliberate anticipation of others and the surplus representations of it entails are neither flattery nor mere accommodation. It is not about pandering to imagined audiences or diluting claims to secure approval. Rather, to embed possible otherness into one's judgment is to adopt an epistemic stance grounded in a fundamental recognition: we do not think in isolation but inhabit a world already populated by others—each with their own perspectives, histories, and situated differences. As Wittgenstein and Davidson long ago emphasized, meaning and truth are inherently social phenomena. Language gains sense only within forms of life sustained by shared practices; truth finds its foothold not in private certainty, but in the dynamic triangulation of speaker, interpreter, and world. Neither meaning nor objectivity can take root in the sealed chamber of individual consciousness. They emerge instead through responsive dialogue, mutual accountability, and the continuous negotiation of intelligibility across difference. To ignore this social grounding risks lapsing into solipsism. Without openness to the critical presence of others—even those who are hypothetical or absent—claims may persist as personal convictions, but they cannot become knowledge in the scientific or communal sense. Publicity, then, is not an add-on to inquiry; it is its constitutive horizon.

Today, this insight might carry growing urgency. In an era marked by deepening social fragmentation, surging distrust in institutions, and widespread skepticism—even hostility—toward scientific expertise. In such a climate, the call for "publicity" in science should be heard not as a technical afterthought but as a foundational aspiration. Yet publicity must not be mistaken for mere transparency. Simply posting datasets online, publishing methodological appendices, or making code repositories public does not, by itself, constitute genuine openness if those materials remain inaccessible, unintelligible, or unusable by actual communities of interpreters.

Authentic publicity demands something more demanding—and more human: empathetic representation. This is the imaginative and ethical labor of anticipating how others might encounter one's work—not just as passive recipients but as active interpreters who may question its assumptions, challenge its boundaries, or extend it in unforeseen directions. It involves crafting representations that are formulated in ways that invite response, enable critique, and make space for co-construction. Only through this kind of engaged, responsive communication can science fulfill its deepest promise: not as the exclusive domain of a credentialed few but as a shared human endeavor—rooted in mutual recognition, sustained by collective reason, and oriented toward the flourishing of all.

Moreover, this line of thought reveals another profound cognitive and ethical truth: even as we rightly attend to the richness of individual experience and cultural diversity, our shared humanity furnishes a vital ground for mutual intelligibility. Differences need not imply incommensurability. Indeed, the very possibility of dialogue, scientific or otherwise, presupposes a horizon of commonality: shared forms of perception, inference, concern, and care that allow us to recognize one another as fellow subjects and agents.

In an increasingly polarized and maybe chaotic world, nurturing this capacity for common understanding might be essential. It underpins not only the cumulative progress of science but also our collective ability to confront planetary-scale challenges: climate disruption, pandemic preparedness, technological (e.g., artificial intelligence) ethics, and systemic inequities. These problems demand cooperation across differences, grounded in reciprocal accountability.

References

Agazzi, Evandro. *Scientific Objectivity and Its Contexts*. Cham: Springer, 2014.

Aristotle. *Politics*, 1253b, 1259b. Translated by Ernest Barker, revised by Richard Stalley. Oxford: Oxford University Press, 1995.

Axtell, Guy. *Objectivity*. Cambridge: Polity Press, 2016.

Bell, John. "Hermann Weyl's later philosophical views: his divergence from Husserl." In *Husserl and the Sciences*. Edited by Richard Feist, 173–185. Ottawa: University of Ottawa Press, 2004.

Burge, Tyler. *Origins of Objectivity*. Oxford: Clarendon Press, 2010.

Cartwright, Nancy, et al. *The Tangle of Science: Reliability Beyond Method, Rigour, and Objectivity*. Oxford: Oxford University Press, 2022.

Carnegy-Arbuthnott, Hannah. "Privacy, publicity, and the right to be forgotten." *Journal of Political Philosophy*, 31 (September, 2023): 494–516.

Crossley, Nick. *Intersubjectivity: The Fabric of Social Becoming*. London: Sage Publications, 1996.

Daston, Lorraine and Peter Galison. *Objectivity*. Cambridge, MA: MIT Press, 2007.

Davidson, Donald. *Subjective, Intersubjective, Objective*. Oxford: Clarendon Press, 2001.

Davidson, Donald. "Three varieties of knowledge." In *Subjective, Intersubjective, Objective*. Oxford: Clarendon Press, 2001.

Davidson, Donald. "The second person." In *Subjective, Intersubjective, Objective*, 107–121. Oxford: Clarendon Press, 2001.

Davidson, Donald. "Externalisms." In *Interpreting Davidson*. Edited by Petr Kotatko, Peter Pagin, and Gabriel Segal, 1–16. Stanford, CA: CSLI Publications, 2001.

Davidson, Donald. "The problem of objectivity." In *Problem of Rationality*. Oxford: Clarendon Press, 2004.

Dear, Peter. "Objectivity in historical perspective." *Metascience*, 21 (January, 2012): 11–39.

Debs, Talal and Michael Redhead. *Objectivity, Invariance, and Convention: Symmetry in Physical Science*. Cambridge, MA: Harvard University Press, 2007.

Douglas, Heather. "Inductive risk and values in science." *Philosophy of Science*, 67 (December, 2000): 559–579.

Douglas, Heather. "The irreducible complexity of objectivity." *Synthese*, 138 (February, 2004): 453–473.

Einstein, Albert. "A Generalized Theory of Gravitation." *Review of Modern Physics*, 20 (January, 1948): 35–39.

Einstein, Albert. "The foundation of the General Theory of Relativity." In *The Principle of Relativity.* Edited by Hendrik Lorentz, et al., 111–164. New York: Dover, 1952.

Feist, Richard. "Husserl and Weyl: phenomenology, mathematics, and physics." In *Husserl and the Sciences*. Edited by Richard Feist, 153–172. Ottawa: University of Ottawa Press, 2004.

Folina, Janet. "Intuition between the analytic-continental divide: Hermann Weyl's philosophy of the continuum." *Philosophia Mathematica*, 16 (2008): 25–55.

Frege, Gottlob. *The Foundations of Arithmetic: A Logico-Mathematical Enquiry into the Concept of Number*. New York: Harper Brothers, 1960.

Goenner, Hubert. "On the history of unified field theories." *Living Reviews in Relativity*, 7 (December, 2004): 2.

Hacking, Ian. *Representing and Intervening: Introductory Topics in the Philosophy of Natural Science*. New York: Cambridge University Press, 1983.

Hacking, Ian. "Let's not talk about objectivity." In *Objectivity in Science: New Perspectives from Science and Technology Studies*. Edited by Flavia Padovani, Alan Richardson, and Thomas Uebel, 19–33. Cham: Springer, 2015.

Harding, Sandra. *Objectivity and Diversity*. Chicago: University of Chicago Press, 2015.

Husserl, Edmund. *The Crisis of European Sciences and Transcendental Phenomenology*. Evanston: Northwestern University Press, 1970.

Husserl, Edmund. *Cartesian Meditations: An Introduction to Phenomenology*. Translated by Dorion Cairns. Hague: Martinus Nijhoff, 1991.

Janack, Marianne. "Dilemmas of objectivity." *Social Epistemology*, 16 (November, 2002): 267–281.

Kant, Immanuel. *Critique of Pure Reason*, A820/B848. Translated by Patricia Kitcher. Indianapolis/Cambridge: Hackett, 1996.

Kitcher, Philip. *The Advancement of Science: Science Without Legend and Objectivity Without Illusions*. New York: Oxford University Press, 1993.

Koskinen, Inkeri. "Objectivity in contexts: withholding epistemic judgement as a strategy for mitigating collective bias." *Synthese*, 199 (April, 2021): 211–225.

Kuhn, Thomas. *The Structure of Scientific Revolutions*. Chicago: University of Chicago Press, 1962.

Kuhn, Thomas. "Objectivity, value judgment, and theory choice." In *The Essential Tension – Selected Studies in Scientific Tradition and Change*. Edited by Thomas Kuhn, 320–339. Chicago: The University of Chicago Press, 1977.

Longino, Helen. *Science as Social Knowledge*. Princeton, NJ: Princeton University Press, 1990.

Longino, Helen. *The Fate of Knowledge*. Princeton, NJ: Princeton University Press, 2002.

Lloyd, Elisabeth. "Objectivity and the double standard for feminist epistemologies." *Synthese*, 104 (September, 1995): 351–381.

Lloyd, Elisabeth. *The Case of the Female Orgasm: Bias in the Science of Evolution*. Cambridge, MA: Harvard University Press, 2005.

McDowell, John. "Subjective, intersubjective, objective." *Philosophy and Phenomenological Research*, 67 (November, 2003): 675–681.

Montminy, Martin. "Triangulation, objectivity and the ambiguity problem." *Crítica: Revista Hispanoamericana de Filosofía*, 35 (December, 2003): 25–48.

Mancosu, Paolo. *The Adventure of Reason: Interplay Between Philosophy of Mathematics and Mathematical Logic, 1900–1940*. New York: Oxford University Press, 2010.

Megill, Allan. "Four senses of objectivity." In *Rethinking Objectivity*. Edited by Allan Megill, 1–29. Durham: Duke University Press, 1994.

Mühlholzer, Felix. "On objectivity." *Erkenntnis*, 28 (1988): 185–230.

Munoe, Paul. "Intersubjectivity." In *The Blackwell Encyclopedia of Sociology*. Edited by George Ritzer and Chris Rojek. Oxford: Wiley-Blackwell, 2019.

Nagel, Thomas. *The View From Nowhere*. New York: Oxford University Press, 1986.

Nissenbaum, Helen. *Privacy in Context: Technology, Policy, and the Integrity of Social Life*. Stanford, CA: Stanford University Press, 2009.

Nozick, Robert. *Invariance: The Structure of the Objective World*. Cambridge, MA: Belknap Press of Harvard University Press, 2001.

O'Raifeartaigh, Lochlainn and Norbert Straumann. "Gauge theory: history, origins and some modern developments." *Reviews of Modern Physics*, 72 (January, 2000): 1–23.

Pauen, Michael. "The second-person perspective." *Inquiry*, 55 (February, 2012): 33–49.

Popper, Karl. *Conjectures and Refutations: The Growth of Scientific Knowledge*. New York: Harper, 1962.

Popper, Karl. *The Logic of Scientific Discovery*. London and New York: Routledge, 2002.

Quine, Willard Van. *The Pursuit of Truth*. Cambridge, MA: Harvard University Press, 1992.

Reich, Wendelin. "Three problems of intersubjectivity—and one solution." *Sociological Theory*, 28 (March, 2010): 40–63.

Reiss, Julian and Jan Sprenger. "Scientific objectivity." In *The Stanford Encyclopedia of Philosophy* (Winter 2020 Edition). Edited by Edward N. Zalta.

Rumbold, Benedict and James Wilson. "Privacy rights and public information." *Journal of Political Philosophy*, 27 (March, 2019): 3–25.

Ryckman, Thomas. *The Reign of Relativity*. New York: Oxford University Press, 2005.

Ryckman, Thomas. "Hermann Weyl, the Gauge principle, and symbolic construction from the 'Purely Infinitesimal'." In *One Hundred Years of Gauge Theory*. Edited by Silvia De Bianchi and Claus Kiefer, 161–183. Switzerland: Springer, 2020.

Sieroka, Norman. "Neighbourhoods and intersubjectivity: analogies between Weyl's analyses of the continuum and transcendental-phenomenological theories of subjectivity." In *Weyl and the Problem of Space*. Edited by Julien Bernard and Carlos Lobo, 99–122. Switzerland: Springer, 2019.

Stephen, John. *Objectivity in Science*. Cambridge: Cambridge University Press, 2021.

Talmage, Catherine. "Meaning and triangulation." *Linguistics and Philosophy*, 20 (April, 1997): 139–145.

Tieszen, Richard. "The philosophical background of Weyl's mathematical constructivism." *Philosophia Mathematica*, 8 (2000): 274–301.

Tsou, Jonathan, Alan Richardson, and Flavia Padovani. "Introduction: objectivity in science." In *Objectivity in Science: New Perspectives From Science and Technology Studies*. Edited by Flavia Padovani, Alan Richardson, and Thomas Uebel, 1–15. Cham: Springer International Publishing, 2015.

Verheggen, Claudine. "Triangulation." In *A Companion to Donald Davidson*. Edited by Ernie Lepore and Kirk Ludwig, 456–472. Oxford: Wiley Blackwell, 2013.

Weyl, Hermann. "Geometrie und Physik." *Die Naturwissenschaften*, 19 (1931): 49–58.

Weyl, Hermann. *Philosophy of Mathematics and Natural Science*. Princeton, NJ: Princeton University Press, 1949.

Weyl, Hermann. *Space-Time-Matter*. New York: Dover Publications, 1952.

Weyl, Hermann. *The Continuum*. New York: Dover Publications, 1994.

Weyl, Hermann. "Mind and nature." In *Mind and Nature: Selected Writings on Philosophy, Mathematics, and Physics*. Edited by Peter Pesic, 83–161. Princeton, NJ and Oxford: Princeton University Press, 2009.

Weyl, Hermann. "Insight and reflection." In *Mind and Nature: Selected Writings on Philosophy, Mathematics, and Physics*. Edited by Peter Pesic, 204–221. Princeton, NJ and Oxford: Princeton University Press, 2009.

Weyl, Hermann. "The open world." In *Mind and Nature: Selected Writings on Philosophy, Mathematics, and Physics*. Edited by Peter Pesic, 34–82, 175–193. Princeton, NJ and Oxford: Princeton University Press, 2009.

Weyl, Hermann. "Man and foundation of science." In *Mind and Nature: Selected Writings on Philosophy, Mathematics, and Physics*. Edited by Peter Pesic, 175–193. Princeton, NJ and Oxford: Princeton University Press, 2009

Wittgenstein, Ludwig. *Philosophical Investigations* (4th ed.). Translated by Gertrude Anscombe, Peter Hacker, and Joachim Schulte. Oxford: Wiley-Blackwell, 2009.

Index

Note: *Italicized* page references refer to figures, **bold** references refer to tables, and page references with "n" refer to endnotes.

For Product Safety Concerns and Information please contact our EU representative GPSR@taylorandfrancis.com Taylor & Francis Verlag GmbH, Kaufingerstraße 24, 80331 München, Germany

Printed and bound by CPI Group (UK) Ltd, Croydon, CR0 4YY
14/09/2026
02216243-0011